Options for Meeting the Maintenance Demands of Active Associate Flying Units

John G. Drew, Kristin F. Lynch, James M. Masters,
Robert S. Tripp, Charles Robert Roll, Jr.

Prepared for the United States Air Force
Approved for public release; distribution unlimited

The research described in this report was sponsored by the United States Air Force under Contracts F49642-01-C-0003 and FA7014-06-C-0001. Further information may be obtained from the Strategic Planning Division, Directorate of Plans, Hq USAF.

Library of Congress Cataloging-in-Publication Data

Options for meeting the maintenance demands of active associate flying units / John G. Drew ... [et al.].
p. cm.
Includes bibliographical references.
ISBN 978-0-8330-4210-1 (pbk. : alk. paper)
1. Airplanes, Military—United States—Maintenance and repair. 2. United States—Air National Guard 3. United States. Air Force—Operational readiness. 4. Air pilots, Military—United States.

UG1243.O68 2008
358.4'162—dc22

2007050674

Cover photo courtesy of the 180 FW, Ohio ANG

Published 2008 by the RAND Corporation
1776 Main Street, P.O. Box 2138, Santa Monica, CA 90407-2138
1200 South Hayes Street, Arlington, VA 22202-5050
4570 Fifth Avenue, Suite 600, Pittsburgh, PA 15213-2665
RAND URL: http://www.rand.org/
To order RAND documents or to obtain additional information, contact
Distribution Services: Telephone: (310) 451-7002;
Fax: (310) 451-6915; Email: order@rand.org

Preface

As the U.S. Air Force faces active component manpower end strength reductions,[1] it becomes more difficult to support the air and space expeditionary force (AEF) construct using current force employment practices. Active component pilot production goals that are already difficult to meet will become even more difficult to achieve in the future.

The Air National Guard (ANG) will retire a significant number of legacy aircraft and realign the remaining aircraft, leading to an increase of primary assigned aircraft (PAA) at most ANG F-16 units (in support of the Quadrennial Defense Review [QDR] and base realignment and closure [BRAC]).[2] At the same time, F-16 units in the active component face challenges in providing adequate flying hours to train their pilots while sustaining an experienced maintenance force. Active associate units[3]—in which active component personnel *associate,* or work, with an ANG unit—could help the Air Force maintain pilot production levels as the ANG offers access to aircraft and to experienced and seasoned pilots to help in training active component pilots. Moreover, these units could help relieve some of the burden on the active component to train and season maintenance personnel.

---

[1] U.S. Air Force, Program Budget Decision 720, December 2005b.

[2] U.S. Department of Defense, "BRAC Commission Actions," briefing, September 1, 2005; U.S. Department of Defense, *Quadrennial Defense Review Report,* September 30, 2001. For example, the BRAC Commission calls for the elimination of the flying mission of a number of ANG flying units operating the A-10, F-16, C-130, and C-135 aircraft.

[3] Also called reverse associate units.

RAND Project AIR FORCE (PAF) was asked to evaluate maintenance organizations for associate units. The analysis is divided into two parts. The first concentrates on understanding and explaining the standards-based differences in aircraft maintenance productivity between active duty and ANG units. The research includes a methodology to quantify and compare the key factors that allow the ANG to generate peacetime training sorties with a fairly small full-time workforce. The second part of the analysis uses the key factors to establish staffing options for an active associate unit, the goal of which is to produce trained pilots in the most efficient manner possible. The analysis shows how various types of personnel can influence the size and productivity of the future workforce in an associate unit.

This monograph is intended to help inform force planning decisions, including those associated with QDR, BRAC, and the Total Force Integration (TFI) effort. Further manpower reductions only heighten the need for a continued review of roles and missions within the different components of the Total Force.

The Director of the ANG Bureau sponsored this research, which was conducted in the Resource Management Program of PAF as part of a project entitled "Evaluation of Air National Guard Transformation Options." The research for this monograph was completed in September 2006.

This monograph should be of interest to maintenance personnel, operators, and force planners throughout the Department of Defense, especially those in the ANG and the active duty Air Force. We assume that the reader has some knowledge of or experience with basic issues surrounding pilot training and aircraft maintenance.

This monograph is one of a series of RAND reports that address agile combat support issues in implementing the AEF. Other related publications include the following:

- *Supporting Expeditionary Aerospace Forces: An Analysis of F-15 Avionics Options*, by Eric Peltz, Hyman L. Shulman, Robert S. Tripp, Timothy Ramey, and John G. Drew (MR-1174-AF, 2001). This report examines alternatives for meeting F-15 avionics maintenance requirements across a range of likely scenarios.

- *Supporting Expeditionary Aerospace Forces: Expanded Analysis of LANTIRN Options*, by Amatzia Feinberg, Hyman L. Shulman, Louis W. Miller, and Robert S. Tripp (MR-1225-AF, 2001). The authors evaluate alternatives for meeting Low Altitude Navigation and Targeting Infrared for Night (LANTIRN) support requirements for AEF operations.
- *Supporting Expeditionary Aerospace Forces: Alternatives for Jet Engine Intermediate Maintenance*, by Mahyar A. Amouzegar, Lionel A. Galway, and Amanda B. Geller (MR-1431-AF, 2002). This report evaluates the manner in which jet engine intermediate maintenance (JEIM) shops can best be configured to facilitate overseas deployments.
- *Supporting Air and Space Expeditionary Forces: Analysis of Maintenance Forward Support Location Operations*, by Amanda Geller, David George, Robert S. Tripp, Mahyar A. Amouzegar, and Charles Robert Roll, Jr. (MG-151-AF, 2004). This monograph discusses the conceptual development and recent implementation of maintenance forward-support locations (also known as centralized intermediate repair facilities [CIRFs]) for the U.S. Air Force.
- *Strategic Analysis of Air National Guard Combat Support and Reachback Functions*, by Robert S. Tripp, Kristin F. Lynch, Ronald G. McGarvey, Don Snyder, Raymond A. Pyles, William A. Williams, and Charles Robert Roll, Jr. (MG-375-AF, 2006). The authors analyze transformational options for better meeting combat support mission needs for the AEF. The role that the ANG may play in these transformational options is evaluated.
- *Supporting the Future Total Force: A Methodology for Evaluating Potential Air National Guard Mission Assignments*, by Kristin F. Lynch, John G. Drew, Sally Sleeper, William A. Williams, James M. Masters, Louis Luangkesorn, Robert S. Tripp, Dahlia S. Lichter, and Charles Robert Roll, Jr. (MG-539-AF, 2007). This monograph develops a methodology that can be used to evaluate potential support posture options for the Future Total Force employing the ANG.

RAND Project AIR FORCE

RAND Project AIR FORCE (PAF), a division of the RAND Corporation, is the U.S. Air Force's federally funded research and development center for studies and analyses. PAF provides the Air Force with independent analyses of policy alternatives affecting the development, employment, combat readiness, and support of current and future aerospace forces. Research is conducted in four programs: Aerospace Force Development; Manpower, Personnel, and Training; Resource Management; and Strategy and Doctrine.

Additional information about PAF is available on our Web site: http://www.rand.org/paf

Contents

Figures

Tables

Summary

As the Air Force faces end strength reductions and force structure changes, it becomes more difficult to support the AEF construct using current force employment practices. To meet congressionally mandated end strength ceilings, the Air Force must eliminate approximately 40,000 active duty personnel in the next several years, without sacrificing operational capabilities. If the Air Force desires to keep pilot production at or near 1,000 pilots per year,[1] alternative organizational structures and resource utilization need to be considered. One of these alternative solutions is to use associate units[2] of the highly experienced ANG workforce and the increased PAA per ANG unit (as a result of the QDR and BRAC decisions) to relieve some of the burden of active component pilot training. With that goal in mind, PAF was asked by senior leaders, both in the ANG and on the Air Staff,[3] to evaluate asso-

[1] The Four Star Summit in 1996 set Air Force pilot production goals at 1,100 (total) and 370 for fighters. Since that time, the fighter goal was slightly reduced and shared with the reserve component (Four Star Summit in April 1999). At the 2003 CORONA, both production goals were reduced by approximately 10 percent. *Transformational Aircrew Management for the 21st Century Tactical Communication Plan* currently lists 1,000 plus or minus 5 percent as the annual Air Force pilot production goal (U.S. Air Force, *Transformational Aircrew Management for the 21st Century Tactical Communication Plan,* May 15, 2007).

[2] An *active* associate unit, also called a reverse associate unit, is an ANG (or Air Force Reserve) unit in which a cadre of active component personnel is permanently assigned to *associate,* or work, with the reserve component unit at the reserve component unit's location.

[3] This analysis was requested by the director of the ANG Bureau and the Directorate of Total Force Integration (AF/A8F) and supported by the active and reserve components' senior staff.

ciate unit maintenance organizations, which could be used to train junior maintenance personnel and to help relieve the burden of active component pilot training.

The research in this monograph focuses on options for how best to meet the requirements for active associate unit aircraft maintenance if some of the active component pilot training requirements were transferred to the ANG. The analysis is divided into two parts. The first concentrates on understanding the differences between ANG and active component aircraft maintenance productivity. The second part uses the key factors to establish staffing options for an active associate unit in which the goal of the unit is to produce trained pilots in the most efficient manner possible. To understand the staffing requirements, a model is used to determine whether a second shift would be required at an active associate unit.

Past RAND analyses found that an ANG unit is able to generate its peacetime training sorties with a fairly small full-time workforce[4]—about one-third the size of the traditional active component organization. Table S.1 compares the total programmed flying hours per full-time maintainer of all Air Combat Command (ACC) F-16 bases with those of all ANG F-16 bases.

Key Factors in the Differences in Productivity

The first part of the analysis focuses on understanding productivity differences (see pp. 7–30). Using past research and discussions with key personnel, we derived the following list of potential key factors:

---

[4] Robert S. Tripp, Kristin F. Lynch, Ronald G. McGarvey, Don Snyder, Raymond A. Pyles, William A. Williams, and Charles Robert Roll, Jr., *Strategic Analysis of Air National Guard Combat Support and Reachback Functions,* Santa Monica, Calif.: RAND Corporation, MG-375-AF, 2006; and Kristin F. Lynch, John G. Drew, Sally Sleeper, William A. Williams, James M. Masters, Louis Luangkesorn, Robert S. Tripp, Dahlia S. Lichter, and Charles Robert Roll, Jr., *Supporting the Future Total Force: A Methodology for Evaluating Potential Air National Guard Mission Assignments,* Santa Monica, Calif.: RAND Corporation, MG-539-AF, 2007.

Table S.1
Comparison of Active Component and ANG F-16 Programmed Flying Hours per Full-Time Equivalent Maintenance Authorization for Fiscal Year 2005

	Combat Coded F-16 Units	
	ACC	ANG
PAA[a]	198	291
Programmed flying hours (PFH)[b]	53,222	76,586
Full-time authorizations[c]	5,629	3,039
Part-time authorizations	0	5,201
Total authorizations	5,629	8,240
Full-time equivalents (FTE)	5,629	3,559
PFH/FTE	9.5	21.5

[a] Manpower data are based on authorizations, not actual fill rates. PAA data for ACC and ANG are from Air Combat Command, Directorate of Logistics, Maintenance Analysis Division.

[b] Programmed flying hour data are from U.S. Air Force, *Air Combat Command, Directorate of Maintenance and Logistics, Ten Year Lookback Standards and Performance FY96–FY05,* HQ Air Combat Command, Directorate of Maintenance and Logistics, December 2005a, and the ANG, Director of Logistics, (ANG/LG).

[c] Full-time authorizations data are from U.S. Air Force, Directorate of Maintenance, Base Level Policy Division (AF/A4MM).

(1) wartime versus peacetime manning factors; (2) out-of-hide duties[5]; (3) on-the-job training (OJT) requirements; (4) supervisory policies; (5) shift or scheduling and utilization efficiencies; (6) depth and range of experience and cross-utilization; and (7) personnel availability. For example, we might expect to see a difference in peacetime productiv-

[5] Out-of-hide responsibilities are those duties that are performed by a maintainer but are not earned through a Logistics Composite Model allotment—for example, the squadron resources manager, squadron small computer manager, dormitory manager, squadron safety noncommissioned officer (NCO), and squadron mobility NCO.

ity, because unit maintenance manpower is sized for wartime flying requirements, which are significantly higher than the unit's peacetime flying requirements. A review of active component flying activity indicates that active duty units may, in fact, be working at or near their full capacity. Therefore, wartime versus peacetime staffing policies do not account for the standards-based differences in ANG and active component productivity, and we do not consider them to be a key factor.[6] Table S.2 summarizes other possible key factors that may contribute to the standards-based differences between ANG and active component peacetime-training sortie generation.

Assessments were developed based on detailed empirical data and expert judgments to quantify the relative effect of each of these factors on a unit's productivity. Figure S.1 illustrates the relative importance of each of the key factors that influence maintenance productivity in an active component maintenance unit. Based on the analysis presented in this monograph (see Chapter Two), a typical active component unit experiences approximately 47 percent of its maximum potential effectiveness per assigned person. In comparison, an ANG unit achieves approximately 90 percent effectiveness per person. The difference in effectiveness between the ANG and active duty units is directly attributable to distinctions in training burdens, availability of manpower, experience levels, and related management practices (see Figure S.1). If these key factors were equal for the active component, the active component and ANG units' net effectiveness could be similar.

While the focus of this study is on F-16 aircraft maintenance, many of these factors (for example, out-of-hide duties, OJT, depth and range of experience, and personnel availability) could affect productivity in other mission areas as well. The relative value of the factors may differ among mission areas, but the factors themselves could influence productivity of other Air Force operations.

---

[6] The effects of split operations (in which part of a unit is deployed forward and part of the unit remains in the rear) and fill rates (assigned personnel versus authorized personnel) are not captured in this analysis. Units may be authorized a certain number of maintainers, but the fill rate could be much lower or have a higher percentage of trainees. Both split operations and the fill rate could affect the results of this analysis.

Table S.2
Comparison of Active Component and ANG Maintenance Organizations

Factor	Active Duty Units	ANG Units
Out-of-hide duties	5% of authorized slots	Negligible
OJT requirements	20% are trainees and are only 40% productive; trainers are 85% productive[a]	Negligible
Supervisory policies	E-7, E-8, and E-9[b] are full-time supervisors	Most supervisors also perform maintenance
Shifts or scheduling	Most maintenance functions run two full shifts[c]	Single-shift maintenance
Depth and range of experience and cross-utilization	Typical enlisted maintainer has 7 years' experience	A typical enlisted maintainer has about 15 years' experience
Personnel availability	Enlisted maintainers spend two days per month on unit training	Full-time techs complete unit training during unit-training assembly (drill weekend) and use leave to do annual training

[a] Steven A. Oliver, *Cost and Valuation of Air Force Aircraft Maintenance Personnel Study,* Maxwell AFB, Gunter Annex, Ala: Air Force Logistics Management Agency, August 2001; Mark J. Albrecht, *Labor Substitution in the Military Environment: Implications for Enlisted Force Management,* Santa Monica, Calif.: RAND Corporation, R-2330-MRAL, 1979; and Carl J. Dahlman, Robert Kerchner, David E. Thaler, *Setting Requirements for Maintenance Manpower in the U.S. Air Force,* Santa Monica, Calif.: RAND Corporation, MR-1436-AF, 2002.

[b] Master sergeants, senior master sergeants, and chief master sergeants.

[c] U.S. Air Force, 2004, authorizes three-shift maintenance at active duty locations.

Figure S.1
Effects of Key Productivity Factors at an Active Component Maintenance Unit

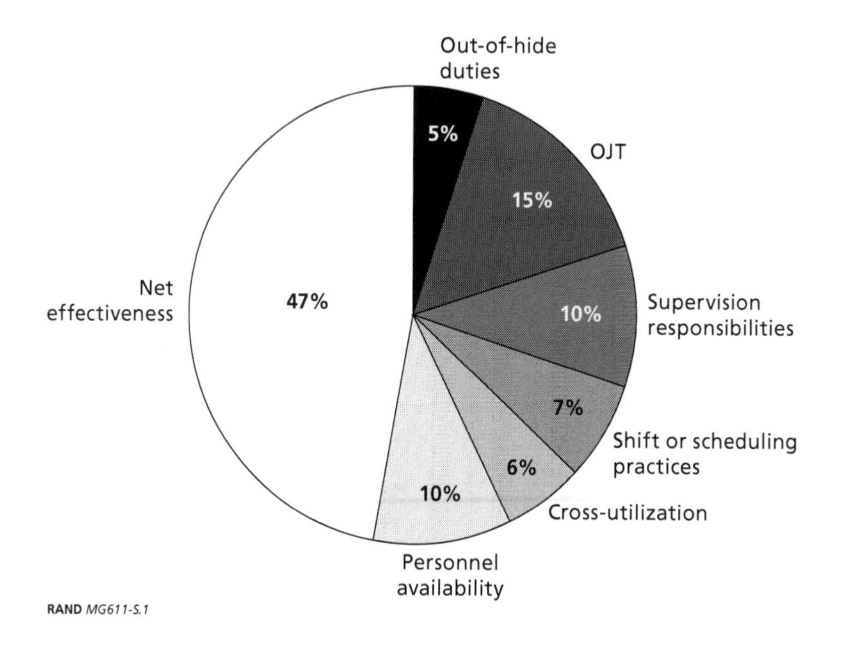

Evaluating Active Associate Maintenance Organization Options

To evaluate options for meeting active associate maintenance requirements, computer simulation models and rules-based applications, which were developed for this analysis, were used to model the flying program and shift operations (see pp. 31–42). Supporting the training of additional pilots requires providing additional training sorties, which would increase the aircraft utilization (UTE) rate. Before active associate staffing requirements could be evaluated, we needed to understand the flying program and how increased maintenance workload might drive a two-shift operation.

To evaluate staffing requirements for an active associate unit, the unit tasking scenario should include the additional PAA added by BRAC, the increased UTE rate in support of TFI requirements, and the personnel impact of a second maintenance shift should it prove necessary (see Table S.3). Taking these into consideration, a unit may need to increase by 45 personnel to run a second shift, which may be required with an increased UTE rate.

Table S.3
Increased PAA and UTE Rate Effects on the Active Associate F-16 Unit

		TFI Scenarios	
	Current	One Shift	Two Shifts
Unit metrics			
PAA	15	18	18
UTE rate[a]	15	18.4	18.4
Average sortie duration (ASD)	1.3	1.3	1.3
PFH (or PAA × UTE rate × ASD)	3,510	5,167	5,167
Manpower authorizations			
Staff	19	22	26
Aircraft maintenance squadron	56	66	68
Maintenance Group leadership	4	4	5
Equipment maintenance squadron	26	33	33
Component maintenance squadron	41	53	59
Total	146	178	191
TFI UTE rate Δ		32	32
Second shift Δ			13
Total increment		32	45

NOTES: The "current" column does not represent any specific unit. Rather, it is a generalized view of what an ANG unit with 15 PAA is accomplishing today. The manpower breakout, however, is closely modeled after the 180 Fighter Wing (FW), Toledo Air Guard Station.

[a] UTE rates are calculated based on crew ratios and pilot training requirements. Pilot training requirements are defined by the Ready Aircrew Program, which differs for each Air Force component—ANG, the reserves, and the active component.

Summary of Findings

There are several factors that contribute to the ANG F-16 unit generating more peacetime flying hours per FTE maintainer than does an active component F-16 unit. First, the ANG, by its very nature, is made up of units possessing a highly experienced workforce. Historically, ANG members remain in the same location much longer (with an average of 15 years experience) than their active component counterparts (with an average of 7 years experience), which deepens their knowledge. Because of their extensive knowledge, the ANG is able to cross-train many of its personnel. Second, the traditional ANG unit recruits its full-time maintenance force (technicians) from a pool of fully qualified applicants. Thus, the unit is able to spend more time on direct production tasks and less time performing initial or upgrade maintenance training than a comparable active unit does. The active component, on the other hand, has a large number of inexperienced maintainers who require hands-on maintenance training and supervision. Active component personnel also have other military duties that reduce their relative availability to perform hands-on maintenance. Finally, the typical active component unit operates two maintenance shifts per day. While two shifts can make a unit very effective, this schedule is inherently less efficient. Most ANG units operate only a single maintenance shift to support peacetime flying sortie generation.

The methodology developed in this monograph can be used to quantify and compare the key factors that allow the ANG to generate peacetime pilot training sorties with a fairly small full-time technician workforce. By applying the methodology to proposed future operations and the proposed TFI associate unit initiatives, this approach can demonstrate how various types of personnel can influence the size and productivity of a proposed unit. If the focus of the TFI initiatives is to improve efficiency, use of full-time ANG maintainers to provide peacetime training sorties for active component pilots may be a viable solution.

Acknowledgments

Numerous persons inside and outside of the Air National Guard provided valuable assistance and support to our work. We thank Lt. Gen. Daniel James III, Director, Air National Guard, for supporting this analysis and Lt. Gen. Craig McKinley for continuing it. We also thank Brig. Gen. David Brubaker, Deputy Director, Air National Guard, and Brig. Gen. Charles Ickes, Chief Operations Officer, Air National Guard, for their support of this effort.

We are especially grateful for the assistance given to us by Air National Guard Brig. Gen. Duane Lodrige, Director ANG Future Total Force; Brig. Gen. Tom Lynn, Director of Logistics; and Rich Rico, Deputy Director of Logistics. Brig. Gens. Lodrige and Lynn and Mr. Rico provided free and open access to everyone on their staffs during our analysis. All were invaluable in providing information and points of contact both inside and outside the Air National Guard.

At the Total Force Integration Office, we thank Maj. Gen. Patrick Gallagher, Brig. Gen. Allison Hickey, Lt. Col. James Godwin, and Maj. Tammy Cobb. At ACC, we thank Lt. Col. Joe Speckhart and Dan Swaney, Directorate of Plans and Programs, Basing Division (ACC/A5B) and Dick Stochetti, Air Combat Command, Directorate of Logistics, Maintenance Analysis Division. At Air Mobility Command, we thank Col. Keith Keck and Maj. Todd Wright, AMC, Directorate of Plans and Programs, Strategy Section. We thank Col. Mike Walters, former maintenance group commander, and Lt. Col. Cheryl Minto, former component maintenance squadron commander, 20 FW, Shaw Air Force Base, South Carolina. At the 180 FW, Ohio ANG, we thank Lt. Col. Jim Reagan, CMSgt Bill Gummow, CMSgt Jim Duty, CMSgt Scott Boyer, CMSgt Claudia Jones, SMSgt Tim Boros, and MSgt Mike Dickman.

At the Air Staff, we thank Col. Dave Whipple and Col. John Stankowski, Directorate of Maintenance, Weapon Systems Sustainment Division (AF/A4MY); Col. Bruce Schmidt, Lt. Col. Dennis Dabney, CMSgt Fred McGregor, CMSgt Elsworth Brown, and Matt McMahan, Directorate of Maintenance, Maintenance Management Division (AF/A4MM).

And finally, at RAND, we thank John Ausink, Robert Kerchner, Dahlia Lichter, Louis Luangkesorn, Tom Manacapilli, Sally Sleeper, Bill Taylor, and Skip Williams. We also thank Bill Taylor and Leslie Lunger at the ACC office for RAND for their continuing support. We would especially like to thank Isaac Porche and David Thaler for their thorough review of this monograph. Their reviews helped shape this document into its final, improved form.

Abbreviations and Acronyms

ACC/A4MQ	Air Combat Command, Directorate of Logistics, Maintenance Analysis Division
ACC	Air Combat Command
AEF	Air and Space Expeditionary Force
AETC	Air Education and Training Command
AF/A4MM	U.S. Air Force, Directorate of Maintenance, Base Level Policy Division
AF/A4MW	U.S. Air Force, Directorate of Maintenance, Weapons and Munitions Division
AF/A8F	U.S. Air Force, Directorate of Total Force Integration
AF/A8FD	U.S. Air Force, Directorate of Total Force Integration, Mission Development Division
AFB	Air Force base
AFRC	Air Force Reserve Command
ARS	Air Reserve Station
AFSO21	Air Force Smart Ops for the 21st Century
AFSOC	Air Force Special Operations Command
AGS	air guard station

AMX	aircraft maintenance squadron
ANG	Air National Guard
ANGB/A4	Air National Guard Bureau, Director of Logistics
ARB	Air Reserve Base
ARC	Air Reserve Component
ASD	average sortie duration
ASOS	air support operations squadron
BRAC	base realignment and closure
CIRF	centralized intermediate repair facility
CMS	component maintenance squadron
CSAF	Chief of Staff of the Air Force
DGS	distributed ground station
E-7	master sergeant
E-8	senior master sergeant
E-9	chief master sergeant
ECM	electronic countermeasures
EMS	equipment maintenance squadron
FOL	forward operating location
FRAP	Fighter Reserve Associate Program
FT	full time
FTE	full-time equivalent
FTU	formal training unit
FW	fighter wing
FY	fiscal year

HQ	headquarters
IAP	international airport
ILM	intermediate-level maintenance
JCA	joint cargo aircraft
JEIM	jet engine intermediate maintenance
JRB	joint reserve base
JSF	Joint Strike Fighter
JSPOC	Joint Space Operations Center
LCOM	Logistics Composite Model
LG	Director of Logistics
MAP	metro airport
MCO	major contingency operation
MRP	Materials Requirement Planning
NAS	naval air station
NASIC	National Air and Space Intelligence Center
NCO	noncommissioned officer
OJT	on-the-job training
PAA	primary assigned aircraft
PAF	RAND Project AIR FORCE
PBD 720	Program Budget Decision 720
PFH	programmed flying hours
PT	part time
QDR	Quadrennial Defense Review

RAIDRS	Rapid Attack, Identification, Detection, and Reporting System
SBIRS MCS	space-based infrared system mission control station
TAMI-21	Transformational Aircrew Management Initiatives for the 21st Century
TFI	Total Force Integration
UAS	unmanned aerial system
UTA	unit training assembly (drill)
UTE	utilization
WFHQ	war fighting headquarters

CHAPTER ONE

Introduction and Research Motivation

As the U.S. Air Force faces end strength reductions and force structure changes required by recent Department of Defense decisions, it becomes more difficult to support the air and space expeditionary force (AEF) construct—a tailored, sustainable force able to respond quickly to national security interests, as needed—using current force employment practices. The Air Force continues to strive to align the Total Force with its primary function—that is, to organize, train, and equip aviation forces primarily for prompt and sustained offensive and defensive air operations[1]—in the most effective way possible with available resources. However, without Air Force action, the end strength and force structure changes will constrain the ability of the Air Force to sustain needed levels of pilot production, especially in the active component. One major initiative the Air Force is proposing to meet this challenge centers on the active associate unit.

Active associate units (sometimes called reverse associate units) could help the Air Force maintain pilot production levels as the ANG offers access to aircraft and to experienced and seasoned pilots to help in training active component pilots. RAND Project AIR FORCE (PAF) was asked by senior leaders, both in the Air National Guard (ANG) and on the Air Staff,[2] to evaluate associate unit maintenance organizations. The research in this monograph focuses on options for

---

[1] U.S. Air Force, Air Force Basic Doctrine, Document 1, November 17, 2003a, p. 43.

[2] This analysis was requested by both the director of the ANG Bureau and the Directorate of Total Force Integration (AF/A8F) and supported by the active and reserve components' senior staff.

how best to meet the *active* associate unit maintenance requirements if some of the active component pilot training requirements were transferred to the ANG.

Definitions

In an *active* associate unit, active component personnel *associate,* or work, with a reserve component unit at a reserve component location. The reserve component has principal responsibility for the weapon system or systems, which it shares with active component personnel. In contrast, in a classic associate unit, reserve component personnel *associate,* or work, with an active component unit at an active component location (see Table 1.1). The active component retains principal responsibility for the weapon system or systems, which it shares with reserve component personnel.

Research Motivation

To meet congressionally mandated end strength ceilings, the Air Force must eliminate approximately 40,000 active duty personnel in the next several years, without sacrificing the operational capabilities outlined

Table 1.1
Active and Classic Associate Unit Definitions

Unit Name	Owns Weapon System at Own Home Station	Cadre of Personnel Who Join the Existing Unit
Active or reverse associate unit	Reserve unit at reserve base	Active component personnel join, or work with, a reserve unit
Classic associate unit	Active component unit at active component base	Reserve personnel join, or work with, an active component unit

in Department of Defense and Air Force planning guidance.[3] In addition, Program Budget Decision 720 (PBD 720) has mandated further manpower reductions, resulting in the total loss of approximately 57,000 personnel through fiscal year 2009 (FY09).[4] Attrition and manpower savings achieved through base realignment and closure (BRAC) will provide some of these manpower reductions. However, approximately 40,000 manpower positions will be eliminated in the Air Force (primarily in the active component), with approximately 20,000 to be eliminated at the start of FY07.

The maintenance career field consists of a large percentage of the current total active duty authorizations. Of the anticipated 40,000 end strength reduction in manpower, the Air Staff expects a reduction of approximately 9,000 maintainers.[5] This is an approximate 11 percent reduction in the 83,854 personnel in maintenance.[6] Under current force employment practices, these manpower reductions may leave the active component without sufficient manpower authorizations to support current operational requirements.

The ANG, on the other hand, will not undergo a significant manpower reduction as a result of BRAC or PBD 720. However, the ANG will be affected by the Air Force force structure planning under way (in support of the QDR and BRAC) that calls for the retirement of a

---

[3] The new Department of Defense Strategic Planning Guidance for fiscal year 2008 and the latest Quadrennial Defense Review (QDR) focus military capabilities on irregular, catastrophic, and disruptive threats. According to the guidance, military capabilities will ensure homeland defense; deter aggression around the globe and, if deterrence fails, be able to engage in two major contingency operations (MCOs) simultaneously or one MCO and one prolonged and irregular conflict. The guidance centers on defeating terrorism; countering nuclear, biological, and chemical weapons; and dissuading major powers from becoming adversaries.

[4] Based on discussions with U.S. Air Force, Directorate of Maintenance, Base Level Policy Division (AF/A4MM). Originally, the reductions were to be achieved by FY11. However, the schedule was recently accelerated to FY09.

[5] Based on discussions with AF/A4MM and U.S. Air Force, Directorate of Maintenance, Weapons and Munitions Division (AF/A4MW). The former expects a reduction of 6,500 positions and the latter 2,500 positions.

[6] File from Consolidated Manpower Database (CMDB), Headquarters Air Force Command Manpower Data System, U.S. Air Force, Directorate of Manpower, Organization, and Resources, September 30, 2004.

significant number of legacy aircraft and a realignment of remaining aircraft. These changes will lead to an increase in primary assigned aircraft (PAA) at most ANG F-16 units.[7]

As a result of PBD 720 and BRAC realignments and reductions, active component pilot production goals that were difficult to meet in years past will become even more difficult to achieve in the future. The Air Force Smart Ops for the 21st Century (AFSO21) implementation plan is focused on improving existing concepts and processes.[8] Combat support transformation, through lean operations and continuous process improvement, is essential to AFSO21 success. AFSO21 improvements may help with active component pilot production goals, but at best, the Air Force may only be able to maintain pre–PBD 720 production capacity. If the Air Force desires to keep pilot production at or near 1,000 pilots per year,[9] alternative organizational structures and resource utilization need to be considered. One of these alternative proposals is the concept of using the highly experienced ANG pilot force to relieve some of the active component pilot training burden through the use of associate units.

The idea of the *associate unit* has a rich history in the U.S. Air Force. Traditionally, these associations typically involve reserve or ANG units associating or colocating with active component units conducting strategic airlift or tanker operations.[10] The idea of asso-

---

[7] U.S. Department of Defense, "BRAC Commission Actions," briefing, September 1, 2005; U.S. Department of Defense, *Quadrennial Defense Review Report,* September 30, 2001. For example, the BRAC Commission calls for the elimination of the flying mission of a number of ANG flying units operating the A-10, F-16, C-130, and C-135 aircraft.

[8] U.S. Air Force, *Air Force Smart Ops for the 21st Century (AFSO21) Implementation Plan: Enabling Excellence in All We Do,* Headquarters (HQ) U.S. Air Force, Directorate of Innovation and Transformation, draft version 3.0, January 31, 2006.

[9] The Four Star Summit in 1996 set Air Force pilot production goals at 1,100 (total) and 370 for fighters. Since that time, the fighter goal was slightly reduced and shared with the reserve component (Four Star Summit in April 1999). At the 2003 CORONA, both production goals were reduced by approximately 10 percent. *Transformational Aircrew Management for the 21st Century Tactical Communication Plan* currently lists 1,000 plus or minus 5 percent as the annual Air Force pilot production goal (U.S. Air Force, *Transformational Aircrew Management for the 21st Century Tactical Communication Plan,* May 15, 2007).

[10] These units are also called classic units or reserve associate units.

ciating other types of Air Force units was reinvigorated with the establishment of the Future Total Force initiative in 1998, involving both the Air Staff and major commands.[11] This effort is now called Total Force Integration (TFI). TFI is tasked to identify potential changes that would, as the name implies, better integrate the active and reserve components into one total force. Today, because the fighter force structure is decreasing and transitioning, it is believed that an associate unit could help the Air Force maintain pilot production levels of 1,000 new pilots a year with at least 300 new fighter pilots. The ANG offers access to aircraft and to experienced and seasoned pilots.

With these associate units in mind, PAF was asked to develop a method for determining how new associate unit maintenance organizations could be structured and how various assumptions about work rules and unit objectives would affect both the active component and the ANG organizations. This research focuses on how to best provide aircraft maintenance for active associate units to support TFI initiatives.[12]

With that focus in mind, the analysis is divided into two parts. The first concentrates on understanding the differences between ANG[13] and active component aircraft maintenance productivity. The analysis investigates seven key factors that could explain the differences between active component and ANG productivity. The second part of the analysis uses the key factors to establish staffing options for an active associate unit. The goal of the unit is to produce trained pilots in the most efficient manner possible. To understand the staffing requirements, a model is used to determine whether a second shift would be required at an active associate unit.

---

[11] The Future Total Force project was originally established in 1998 to explore potential solutions to some of the Air Force's most pressing problems with recruitment, retention, manning, and the budget.

[12] See Appendix A for a list of TFI initiatives.

[13] We consider both full-time and part-time ANG maintenance personnel, adjusting the number of part-time personnel to full-time equivalents (FTEs) (see Chapter Two for details).

The remainder of this monograph is based on the premise that end strength and force structure decisions will exacerbate the active component training burden for both pilots and maintainers and that the ANG would be able to relieve some of that burden through the use of an associate unit. The authors understand that there are many ongoing discussions about the legalities of using the reserve component to train the active component as well as issues centered on command structures, including U.S. Code Title 10 and Title 32 responsibilities. While these issues may not be easily resolved, they are beyond the scope of this analysis.

Organization of This Monograph

Chapter Two investigates the key factors that help explain the standards-based differences in aircraft maintenance productivity between the active component and the ANG. Chapter Three evaluates options for meeting active associate maintenance requirements. The summary findings are in Chapter Four. Appendix A presents Total Force Integration initiatives. Appendix B presents productivity examples from active component and ANG F-16 bases. And, finally, Appendix C presents a detailed description of the simulation model used in this analysis.

CHAPTER TWO

Understanding Standards-Based Productivity Differences

This chapter focuses on the first part of the analysis, understanding the standards-based differences between ANG and active component aircraft maintenance productivity. Before beginning the investigation into the key factors that could affect productivity, this chapter will provide some background information on how the productivity issue was uncovered.

Research Approach

According to a 2000 RAND report on pilot shortages, the Air Force is facing the largest peacetime pilot shortage in its history, with about half of the shortfall occurring in fighter pilots.[1] That research identifies low experience levels in operational units as one of the main drivers of the shortfall. Unless sorties and flying hours are increased with the associated aircraft utilization (UTE) rate, the problem will continue to worsen each year. Even if additional flying hours could be provided, the shortage of experienced pilots limits a unit's ability to train its inexperienced pilots. Since 2000, the Air Force has taken several steps in

[1] William W. Taylor, S. Craig Moore, and Charles Robert Roll, Jr., *The Air Force Pilot Shortage: A Crisis for Operational Units?* Santa Monica, Calif.: RAND Corporation, MR-1204-AF, 2000.

an attempt to mitigate this problem;[2] however, a pilot shortage and the pilot training burden still remain. One possible solution would be to use highly experienced ANG pilots to train a portion of the inexperienced active duty pilots at active associate units.

Previous RAND research and analyses provide insights into potential solutions and direction for considering the formulation of an active associate unit.[3] Among these, RAND analyses of continental

---

[2] The Total Force Absorption Program was put into place in 2000; however, it has not achieved its goal of producing enough pilots to meet Air Force needs. Transformational Aircrew Management Initiatives for the 21st Century (TAMI-21) was also sponsored by the Chief of Staff of the Air Force with Air Combat Command to address pilot shortages.

[3] Albert A. Robbert, William A. Williams, and Cynthia R. Cook, *Principles for Determining the Air Force Active/Reserve Mix,* Santa Monica, Calif.: RAND Corporation, MR-1091-AF, 1999, produced a seminal report that established a rational basis for determining the absolute and relative size of the reserve component relative to the active component. This analysis provided baseline estimates of constraints on the proportion of active component and reserve component personnel that would be feasible in an active associate unit.

RAND and the Air Force Logistics Management Agency partnered in 2001 to determine the cost and value of a fully trained Air Force maintenance technician (Steven A. Oliver, *Cost and Valuation of Air Force Aircraft Maintenance Personnel Study,* Maxwell AFB, Gunter Annex, Ala.: Air Force Logistics Management Agency, August 2001). The data derived in the Air Force Logistics Management Agency study provided baseline parameters for the trainer and trainee productivity necessary for the maintenance manpower evaluation conducted in this research.

In 2004, RAND evaluated the potential role of the ANG in four Air Force mission areas: civil engineering deployment and sustainment capabilities, continental U.S. centralized intermediate repair facilities (CIRFs), the Force Structure and Cost Estimating Tool—a planning extension to GUARDIAN capabilities, and reachback missions in the air and space operations center (Robert S. Tripp, Kristin F. Lynch, Ronald G. McGarvey, Don Snyder, Raymond A. Pyles, William A. Williams, and Charles Robert Roll, Jr., *Strategic Analysis of Air National Guard Combat Support and Reachback Functions,* Santa Monica, Calif.: RAND Corporation, MG-375-AF, 2006). And finally, in 2005, RAND developed a methodology that can be used to investigate the role that the ANG may play in assuming some of the missions the active component may not be able to fully staff under current manpower constraints (Kristin F. Lynch, John G. Drew, Sally Sleeper, William A. Williams, James M. Masters, Louis Luangkesorn, Robert S. Tripp, Dahlia S. Lichter, and Charles Robert Roll, Jr., *Supporting the Future Total Force: A Methodology for Evaluating Potential Air National Guard Mission Assignments*, Santa Monica, Calif.: RAND Corporation, MG-539-AF, 2007). Such mission areas as Predator operations and support, air mobility command and control, commander of Air Force forces staffing, base-level intermediate maintenance, and intercontinental ballistic missile maintenance were evaluated because they would capitalize on ANG

U.S. centralized intermediate repair facilities (CIRFs)[4] and base-level intermediate maintenance found that ANG F-16 units are able to generate peacetime training sorties while maintaining the required maintenance capability and with a relatively small full-time maintenance workforce.[5] This ANG workforce is about one-third the size of the traditional active component organization and is typically composed of highly experienced, senior noncommissioned officers (NCOs) and a few commissioned officers. These individuals are civil servants working full time during the week—maintaining aircraft and generating peacetime training sorties—who also hold traditional ANG military positions[6] in the unit.

We conducted the present analytic effort to understand and explain causes of the standards-based planned productivity differences between active duty and ANG F-16 units in generating peacetime training sorties, that is, to determine the factors that contribute to the differences in maintenance productivity between the ANG and the active component. We developed an approach to quantify and compare these key factors. Finally, we applied the insights gained to proposed future unit flying operations to show how various personnel types (full-time technicians, highly qualified active duty members, and trainees) and mixes of them can influence the size and productivity of the proposed maintenance workforce in an active associate unit.

---

strengths and provide effective and efficient approaches to achieving the desired operational effects—supporting the AEF construct from a Total Force perspective.

4 The CIRF concept is one in which intermediate-level maintenance (ILM) is consolidated into a small number of relatively large facilities to enable expeditionary operations through an increased efficiency and reduction in deployed footprint (because the ILM shop is not deployed with the unit to the forward operating location [FOL]). The CIRF concept allows for a reduction in deployment time frames, although it requires a dedicated transport commitment to ship broken commodities from the FOLs to the CIRF along with serviceable commodities from the CIRF to the FOLs.

5 Tripp et al., 2006; Lynch et al., 2007.

6 Traditional ANG personnel work part time—one weekend a month and two weeks a year.

Research Focus

This evaluation of active associate units focuses on F-16 units for several reasons. First, the Air Force's production goal of 300 fighter pilots per year[7] from undergraduate pilot training creates a substantial pilot training burden for fighter squadrons to which these new pilots are assigned. Since the F-16 fleet is a considerable portion of the total fighter fleet, F-16 units bear a significant part of the pilot training burden. In addition, a large percentage of the F-16 fleet is assigned to the reserve component, making that fleet a good place to begin evaluating active associate units.

The F-16 is scheduled to be replaced by the Joint Strike Fighter (JSF) F-35, though not in the near future; the F-16 will be flying for at least ten more years. While not a one-for-one replacement, the JSF fleet is planned to be large. Any organizational or process changes that are made to facilitate F-16 pilot production could be applicable to the F-35 in future years.

Finally, the sheer size of the F-16 fleet requires numerous units at active and reserve component bases. The number and size of the bases involved in F-16 operations allow many opportunities for associations between the components.

After the BRAC decisions of 2005, most ANG F-16 units realized an increase in PAA. Many 15 PAA organizations have become 18 PAA wings. With the additional PAA, units may be better able to support increased flying operations, but only if personnel numbers also increase to support such operations. Table 2.1 lists the reserve component bases with F-16 units and their PAA after the BRAC realignment.

---

[7] The Four Star Summit in 1996 set the Air Force fighter pilot production goal at 370. The goal was slightly reduced to 330 (with 30 in ANG and AFRC units) at the Four Star Summit in April 1999. At the 2003 CORONA, the fighter pilot production goal was further reduced by about 10 percent to approximately 300 for the total force. TAMI-21 calls for the number to gradually increase back to the 1,100 total by the end of FY09.

Table 2.1
Reserve Component F-16 Units

Installation	Organization	PAA
Tucson	ANG	61
McEntire	ANG	24
Homestead	AFRC	24
Ft. Worth	AFRC	24
Tulsa	ANG	21
Toledo	ANG	18
Lackland	ANG	18
Ft. Wayne	ANG	18
Fresno	ANG	18
Des Moines	ANG	18
Dannelly Field	ANG	18
Burlington	ANG	18
Buckley	ANG	18
Atlantic City	ANG	18
Andrews	ANG	18
Madison	ANG	15
Kirtland	ANG	15
Joe Foss Field	ANG	15
Duluth	ANG	15

SOURCE: U.S. Department of Defense, 2005, Slide 8.

NOTES: Tucson is a training base providing formal training unit (FTU) capability for the ANG. AFRC is Air Force Reserve Command.

Understanding Standards-Based Differences in Productivity

There are significant productivity differences between ANG[8] and active component units in producing peacetime training sorties. As illustrated in Table 2.2, an ANG F-16 unit is able to generate more than twice as many peacetime flying hours per FTE maintenance authorization as a comparable active component unit. Table 2.2 compares the annual flying hour and maintenance personnel authorization data for all Air Combat Command (ACC) combat-coded F-16 bases with those of all ANG combat-coded F-16 bases. Specific examples for Shaw Air Force

Table 2.2
Comparison of Active Component and ANG F-16 Programmed Flying Hours per FTE Maintenance Authorization for FY05

	Combat Coded F-16 Units	
	ACC	**ANG**
PAA[a]	198	291
Programmed flying hours (PFH)[b]	53,222	76,586
Full-time authorizations[c]	5,629	3,039
Part-time authorizations	0	5,201
Total authorizations	5,629	8,240
FTE	5,629	3,559
PFH / FTE	9.5	21.5

[a] Manpower data are based on authorizations, not actual fill rates. PAA data for ACC and ANG are from ACC/A4MQ.

[b] Programmed flying hour data are from U.S. Air Force, *Air Combat Command, Directorate of Maintenance and Logistics, Ten Year Lookback Standards and Performance FY96–FY05,* HQ Air Combat Command, Directorate of Maintenance and Logistics, December 2005a, and the ANG, Director of Logistics, (ANG/LG).

[c] Full-time authorizations data are from U.S. Air Force, Directorate of Maintenance, Base Level Policy Division (AF/A4MM).

---

[8] An ANG unit includes both full-time and part-time personnel.

Base (AFB), McEntire Air Guard Station (AGS), and Toledo AGS can be found in Appendix B.

In evaluating the annual programmed flying hours per FTE, only the PAA of an ACC or ANG unit are considered. Neither the backup aircraft inventory nor attrition reserve aircraft assigned to a unit are included. Annual programmed flying hours are calculated by multiplying the total PAA times the UTE rate[9] times the average sortie duration (ASD) times 12 months.

The *inputs* side of the productivity ratio is taken to be the total authorized manpower available to the unit based on standard Air Force unit manning documents. In the case of an ANG unit, the manpower numbers are adjusted to account for the part-time participation of traditional guardsmen in normal peacetime activities.[10] This adjustment is not meant to indicate either that the reserve or active component unit is producing at maximum capacity or that all authorized personnel are equally productive. From an aggregate productivity perspective, the point is to compare inputs and outputs, where inputs are total manpower authorizations and outputs are programmed flying hours. In this analysis, the number of personnel assigned and the peacetime sorties produced represent activity during FY05.

To establish the number of FTE maintenance personnel in an ANG F-16 unit, all full-time maintenance technicians are included in the count. In an ANG unit, the part-time workforce also contributes to producing peacetime sorties. Each traditional guardsman works one weekend a month and two weeks a year (24 days of weekend drill plus 14 days of annual training). Traditional guardsmen must perform annual training (for skill development and refinement) and other military requirements during this time. As a result, the traditional guardsman can contribute no more than 50 percent of this time to actual maintenance work in support of generating peacetime sorties; that is,

---

[9] UTE rates are calculated based on crew ratios and pilot training requirements. Pilot training requirements are defined by the Ready Aircrew Program, which differs for each Air Force component—ANG, the reserves, and the active component.

[10] Approximately 10 percent of that of full-time ANG personnel.

no more than 19 of the 38 days.[11] Comparing the 19 days of availability of the part-time ANG personnel to the 193 days of availability of the active duty personnel[12], a part-time ANG maintainer can be considered as available and contributing to the maintenance mission at about 10 percent of a full-time authorization.[13] Adding the full-time authorization (3,039) and the contribution of the part-time workforce (520)[14] gives an FTE of 3,559 ANG personnel. For the purposes of this analysis, we assume that full- and part-time ANG maintainers provide roughly equal productivity when available for maintenance activities.[15] For the active component unit, all authorized personnel are full time and are therefore considered 100 percent FTE.

The number of flying hours generated per maintainer is more than twice as high in an ANG unit. These computations were completed and validated on several individual bases as well as across all of ACC and ANG F-16 units. In nearly every case, the ANG was able to produce more than two times as many peacetime sorties per full-time maintainer as a comparable active component unit.

Key Factors in the Standards-Based Differences

To be able to develop a rational basis for manning active associate unit aircraft maintenance operations, it is essential to understand why such a large difference exists between the active component and the ANG F-16 units in generating peacetime sorties. Many factors have been suggested as possible explanations or causes of the standards-based differences, including

- wartime versus peacetime manning factors
- "out-of-hide" duties

---

[11] Based on discussions with 180 Fighter Wing, Toledo, Ohio.

[12] See the section on Personnel Availability later in this chapter for a complete explanation of active component personnel availability.

[13] $19 \div 193 = 0.098$.

[14] $5{,}201 \times 10\% = 520$.

[15] Every ten part-time or traditional guardsmen equals one FTE (520), and their productivity is equal to the FTE of 3,039.

- on-the-job training (OJT) requirements
- supervisory policies
- scheduling and utilization efficiencies
- depth of experience and cross-utilization
- personnel availability.

Each of these potential key factors may contribute to the overall difference in productivity, with some perhaps more important than others. It would be very useful to identify the measurable contribution, if any, of each of these factors to a unit's total productivity. The measure should allow an assessment of the impact of each key factor on the overall total productivity. This assessment should provide insight into a portfolio of best practices that can be applied to the associate unit.

Wartime Versus Peacetime Manning Factors. ANG and active component units are both expected to be able to produce their wartime sortie generation capability with the manpower they are allotted through the Logistics Composite Model (LCOM). LCOM manpower allotments are based on the number of flying hours a unit is assigned. Applying historical component break rates and repair times[16] and minimum crew sizes, LCOM uses computer simulations to compute the corresponding maintenance wartime manpower requirement. Personnel availability standards and utilization factors are applied to the total labor requirement to establish manpower authorizations. The peacetime (training) flying requirement and the ability of a single unit to operate from two or more locations (split operations) can also be computed using LCOM procedures. However, currently, LCOM is run only for wartime and peacetime operations, not split operations. Units are then staffed to the greatest of all of these requirements—peacetime or wartime operations. LCOM and major command manpower personnel work together to establish authorizations for each unit. These authorizations are then funded, and personnel are assigned to the positions.[17]

Using LCOM, an ANG unit with all of its manpower slots would look almost identical to a like-sized active component unit. However, to

---

[16] LCOM uses the five-digit work unit code for component break and repair times.

[17] Because of budgetary and other constraints, not all authorized positions are funded.

produce peacetime training sorties, an ANG unit uses a much smaller full-time workforce. In ANG units, about 37 percent of the maintenance authorizations are full-time technician positions.[18] The remaining authorizations are traditional guardsmen; that is, they are essentially part-time employees who are available to work approximately one-tenth as much as a full-time employee unless they are activated.[19]

Since the ANG is able to meet peacetime requirements with a smaller full-time workforce, it may appear that active component units are overmanned for peacetime training sortie production. In other words, because the active component is staffed for higher wartime requirements, it may appear that it would be underutilized in the lower operational tempo of peacetime training sortie production. Because wartime mission requirements usually have more demanding requirements than those for peacetime, there may simply be more full-time personnel assigned to an active component unit than are required during peacetime operations.

However, a review of active component flying activity indicates that active duty units do not appear to have the expected slack capacity that would be found with an overmanned or underutilized unit. In fact, the increased wartime capability is expected to come, in large part, from changes in work rules, for example, an increase in time spent on the job or an increase in personnel availability. Expanded capacity is possible because the normal 40-hour work week (five days times 8 hours per day) is extended to 72 hours (six days times 12 hours per day) during wartime operations.[20] In addition, many of the additional duties that require time spent away from the job are eliminated or reduced significantly during times of war.

Furthermore, the active component F-16 fleet did not meet the UTE rate goal in FY05. The FY05 F-16 UTE rate goal was 16.4; how-

---

[18] ANG technicians are full-time civil servants who work during the week to produce training sorties and are also assigned a traditional ANG position in the wing.

[19] Based on the ANG Full-Time Maintenance Model.

[20] U.S. Air Force, *Air Force Instruction 38-201*, December 30, 2003b.

ever, the actual UTE rate flown was 15.2.[21] Many units reported the routine use of overtime, as well as the use of contract maintenance teams, to meet their peacetime maintenance requirements. Maintenance manpower is not the only contributing factor to the failure of the active component in meeting its UTE rate goal. AEF rotations as well as other factors can all contribute. However, not meeting the UTE rate goal may be one indication that the active component unit is working at or near maximum capacity with its allotted personnel.[22]

This is not meant as a criticism of the active component. Rather, the point is that the active component unit may be working at or near its full capability. And therefore, potential underutilization because of wartime versus peacetime staffing policies does not account for the standards-based differences in ANG and active component productivity. That is, there is no strong evidence that ANG units appear to be more productive because active component units are relatively underutilized.

"Out-of-Hide" Duties. Out-of-hide slots are those positions in the unit that involve processes normally accomplished by a maintainer but the position is not earned or recognized through an LCOM allotment. Examples of these jobs include the squadron resources manager, squadron small computer manager, dormitory manager, squadron safety NCO, and squadron mobility NCO. Typically these are all full-time positions in the active component and are treated as additional duties by the ANG. In addition, the wing commander has the authority to reassign maintenance personnel to perform other high-priority jobs that have little or nothing to do with maintenance. The typical active component maintenance unit has approximately 5 percent of the workforce performing out-of-hide duties at any given time.[23] The

---

[21] U.S. Air Force, *Air Combat Command, Directorate of Maintenance and Logistics, Ten Year Lookback Standards and Performance FY96–FY05*, HQ ACC, Directorate of Maintenance and Logistics, (ACC/A4P), December 2005a.

[22] For an analysis of LCOM and the maintenance environment in the field, see Dahlman, Kerchner, and Thaler, 2002.

[23] Air Force Logistics Management Agency, *Analysis of Out-of-Hide Job Requirements Levied on Aircraft Maintenance Units*, Maxwell AFB, Gunter Annex, Ala., October 2005.

ANG, on the other hand, indicates that out-of-hide duties have little or no impact on an ANG unit.[24]

On-the-Job Training Requirements. In an active component maintenance unit, a significant percentage of the total workforce, approximately 20 percent, are junior-level maintenance personnel or *three levels* (trainees).[25] These three levels are not fully productive, and they also decrease the productivity of the experienced maintenance personnel (five and seven levels—trainers) because the trainers need to stop and instruct or explain as they complete the work, thus taking longer to complete the task than if they were working with another fully trained individual.[26] Since 20 percent of the active component workforce is trainees, approximately 20 percent of the workforce is providing maintenance training (five and seven levels). Thus, the productivity of an active component unit suffers.

Several recent studies have examined the effect of trainees on productivity.[27] Based on those past analyses, average trainee productivity is estimated to be 40 percent of that of a fully trained individual for this analysis. Based on the same past analyses, average trainer productivity is estimated to be 85 percent of that of a fully trained individual, someone who is not acting as a trainer and can work full time without training responsibilities. It is further assumed that the ratio of trainers to trainees, in terms of its effect on productivity, is essentially one to one.[28]

---

[24] Based on discussions with 180 Fighter Wing (FW), Toledo, Ohio.

[25] Based on discussions with Directorate of Maintenance, Base Level Policy Division (AF/A4MM).

[26] We call them "trainers," but the five and seven levels are not dedicated trainers. They are just more experienced personnel doing maintenance work that the three levels can watch and learn from in the process of OJT.

[27] Oliver, 2001; Mark J. Albrecht, *Labor Substitution in the Military Environment: Implications for Enlisted Force Management*, Santa Monica, Calif.: RAND Corporation, R-2330-MRAL, 1979; and Dahlman, Kerchner, and Thaler, 2002.

[28] For the purposes of this analysis, we treat trainers and trainees as single individuals and do not look at the cost or benefit of combining one more from either category—for example one trainer training four trainees. While interesting, it was beyond the scope of this analysis.

The ANG, on the other hand, has very few junior-level personnel (trainees) in its full-time technician workforce—in maintenance and in many other career fields.[29] Individuals are hired into the full-time technician workforce based on their ability to complete the work immediately. Any required training (to develop new skills or further refine skills) takes place on drill weekends or during the two-week annual training. Thus, trainer and trainee productivity effects on ANG productivity are negligible.

Supervisory Policies. The ANG uses almost the entire full-time technician workforce in a hands-on capacity. Even the most senior enlisted personnel perform maintenance work and are only part-time supervisors. In contrast, in an active component unit, maintenance technicians who attain the rank of master sergeant (E-7) become full-time managers. They are no longer involved in hands-on maintenance. This practice may in part be driven by the number of trainees—supervisory tasks include scheduling and tracking of appointments, maintaining training records, administration of career development courses,[30] and other workload associated with training. The differences in supervisory practices between the active component and the ANG may affect unit productivity.

Scheduling and Utilization Efficiencies. The active component strives to keep as many aircraft mission capable as possible, putting pressure on maintainers to fix every broken aircraft quickly. The typical active component unit operates two maintenance shifts; some even have a small servicing crew on a third shift. When maintaining two shifts, an overhead contingent and a minimum crew size are required in each shop or specialty area on each shift, leading to inefficient utilization of labor. Because maintenance managers are unable to predict how much labor will be needed and when, they are forced to position enough manpower on each shift to be able to respond to varying workloads. For example, fuel cell maintenance requires a minimum crew size of three. Because fuel cell maintenance requires special cer-

---

[29] Based on discussions with Air National Guard Bureau, Director of Logistics (ANGB/A4) and AF/A4MM.

[30] Self-study guides that must be completed prior to upgrade.

tifications, a minimum of three individuals would need to be assigned to both the first and second shift—regardless of the fuel call maintenance workload. Likewise, ejection seat maintenance—because of work with explosives—requires two highly qualified individuals per shift for this seldom-used shop. While two shifts make a unit very effective, in terms of returning non–mission capable aircraft to fully–mission capable status as quickly as possible, they are simply not as efficient, from a manpower perspective, as accomplishing all maintenance during a single shift.

Most ANG units, on the other hand, operate only a single shift to support peacetime training sorties. Additionally, the ANG tends to repair "to the flying schedule"—they do not focus on fixing everything as quickly as possible; rather the emphasis is on simply insuring that sufficient flyable aircraft are available to support the next period's flying schedule. Other maintenance is performed as time becomes available.

The ANG also has a different philosophy on providing peacetime training sorties. Suppose a unit needed to fly 16 sorties per day to provide peacetime training sorties. The sorties would be generated in two time blocks or waves. The active component would prefer to fly a schedule with decreasing numbers of aircraft committed to the second and subsequent set of sorties. For example, the active component would schedule 10 sorties for the first wave, with two spare aircraft available. Then, the second wave would be 6 sorties. To support this schedule, the maintenance unit must be able to consistently generate at least 12 flyable aircraft. The ANG takes a very different approach. It would typically operate 6 sorties followed by 10. In this way, the ANG unit needs to generate only 8 flyable aircraft at the beginning of the day and could work during its single shift to generate the additional 4 aircraft needed for the second wave.

Depth of Experience and Cross-Utilization. The average Air Force enlisted person has about seven years of total experience. The first three years of a maintainer's career are spent in formal "schoolhouse" training and OJT. The average active component maintainer moves to a new assignment every three years, often changing aircraft platforms in the process. Even if the type of airframe or the type of repair is the same, this continuous rotation of personnel causes teams to be constantly formed and reformed. Moreover, any training of an individual maintainer not completed during the initial assignment must be taken on by the unit at his or her next assignment. An ANG unit, on the other hand, is relatively stable. The average ANG full-time technician has over 15 years' experience and has been at one location most of his or her career.

Because of this depth of experience, the ANG unit is able to achieve efficiencies through cross-utilization of personnel. For example, the active component unit must be manned with enough dedicated flight line mechanics to perform all launch and recovery operations. In the ANG, every full-time technician, in every shop, is also trained and fully qualified to assist in flight line duties when required. For instance, every full-time ANG technician can perform *B-man* or launch-assist duties, and every full-time member is qualified to assist in towing—the movement of aircraft from one location to another by utilizing ground support equipment. Each ANG unit also qualifies its personnel to assist in more than one maintenance shop, based on need and expected workflow. Thus, the ANG unit can handle variations in peak specialist workloads by moving its available, highly experienced people to where the work happens to be. In contrast, the typical active component unit must staff each shop to be able to independently handle its own peak workload. Thus, the ANG unit is able to capitalize on its depth and range of maintenance experience by operating with fewer full-time maintainers. While the active component has encouraged this type of cross training and cross-utilization, the personnel demographics of the active component maintenance workforce, particularly in terms of its much less experienced workforce, limit its ability to fully implement the concept.

Personnel Availability. All active component personnel must complete ancillary training, maintain proficiency training, and perform other military duties, such as honor guard, bay orderly, and other duties that reduce their availability to perform hands-on maintenance. A cautious estimate of the impact of these additional duties and taskings is approximately two days per month per maintainer.[31]

---

[31] Dahlman, Kerchner, and Thaler, 2002; and Oliver, 2001.

ANG personnel do not have to complete all of the military duties that the active component does. In addition, because all ANG technicians also hold traditional positions in the unit, they have a two-day unit training assembly (UTA), a "drill weekend," once a month to fulfill their training requirements. Thus, the individual availability of ANG personnel is much higher than that of those in the active component. Additionally, the full-time technicians routinely use their civil service leave time during their two-week annual training event. While they are on leave, they are performing maintenance work as traditional guardsmen, so their total availability increases.

Table 2.3 summarizes the possible key factors, discussed above, that may contribute to the standards-based differences between ANG

Table 2.3
Differences Between Active Component and ANG Maintenance Organizations

Factor	Active Duty Units	ANG Units
Out–of-hide duties	5% of authorized slots	Negligible
OJT requirements	20% are trainees and are only 40% productive; trainers are 85% productive[a]	Negligible
Supervisory policies	E-7, E-8, and E-9[b] are full-time supervisors	Most supervisors also perform maintenance
Shifts or scheduling	Most maintenance functions run two full shifts[c]	Single-shift maintenance
Depth and range of experience and cross-utilization	Typical enlisted maintainer has 7 years' experience	A typical enlisted maintainer has about 15 years' experience
Personnel availability	Enlisted maintainers spend two days per month on unit training	Full-time techs complete unit training during unit-training assembly (drill weekend) and use leave to do annual training

[a] Oliver, 2001; Albrecht, 1979; and Dahlman, Kerchner, and Thaler, 2002.

[b] Master sergeants, senior master sergeants, and chief master sergeants.

[c] U.S. Air Force, 2004, authorizes three-shift maintenance at active duty locations.

and active component peacetime training sortie generation. With each factor, the table compares the impact on both an active component and an ANG unit.

While the focus of this study is on F-16 aircraft maintenance, many of these factors (for example, out-of-hide duties, OJT, depth and range of experience, and personnel availability) could affect productivity in other mission areas as well. These factors could be considered when evaluating the productivity of other Air Force operations.

Our analysis tried to identify and explain the differences in productivity between ANG and active component units; however, there are several factors that were not included in this work. The effects of split operations, when part of a unit is deployed forward and part of the unit remains in the rear, are not captured in this analysis. Each of the seven key factors could have an even greater effect on the net effectiveness during split operations. For example, with fewer personnel in the unit to task, out-of-hide duties would take more of the time of those personnel who are available to be tasked, making them less effective. During split operations, a unit typically sends more experienced maintainers forward, which can affect the OJT that the unit can do at home, not to mention the production of sorties for pilot training. Supervision and training of junior maintenance personnel take longer when some of the maintenance supervisory and training staff are deployed to a forward location. Moreover, the unit has to catch up on training when the deployed personnel and equipment return home.

In addition, this analysis looks at authorized personnel, not actual manning. Units may be authorized a certain number of maintainers, but the fill rate could be much lower or have a higher percentage of junior maintenance personnel (three levels, as discussed above). With a higher number of maintenance trainees (three levels), OJT may take even longer to complete. Both split operations and fill rate could affect the results of this analysis even more profoundly, which may cause the active component to appear even more inefficient.

Quantifying the Effect of Key Factors on Unit Productivity

To quantify the effect of these factors on unit productivity, our analysis reduces each unit's total authorized FTEs to an appropriate number of effective FTEs. That is, the analysis will align the apparent total workforce with the number of equivalent fully productive individuals that result from the policies and procedures that the unit employs. It is then possible to compare unit productivity. Data from Shaw AFB, South Carolina, and Toledo AGS, Ohio, are used as examples for this analysis.[32]

In the active component, the programmed flying hours for the 72 F-16s at Shaw AFB are 19,354 hours per year.[33] There are 1,981 maintenance personnel assigned to the F-16 wing at Shaw, which include the Air Force specialty codes 2A, 2W, 2R, and 2P for enlisted personnel and 21A for officers. The annual programmed flying hours per maintainer is 9.8 hours per person.[34]

For the ANG, there are 146 full-time and 270 part-time maintainers authorized for the Toledo AGS F-16 unit of 15 PAA. To account for the contribution of the part-time workforce, 10 percent of the part-time total workforce, or 27 personnel, is added.[35] The remainder of their on-duty time is spent in other training (skill development or refinement).

Adding the 27 part-time personnel to the 146 full-time authorizations gives a total FTE workforce of 173 ANG personnel. The annual programmed flying hours for Toledo AGS are 3,400 hours per year.[36]

---

32 The 20 FW, Shaw AFB, South Carolina, has block 50 F-16s. The 180 FW, Toledo AGS, Ohio, has block 42 F-16s. The difference in maintenance organization, processes, and policies between the two different blocks of aircraft are negligible and do not affect the differences in productivity.

33 PAA times UTE rate times ASD times 12 months equals $72 \times 1.4 \times 16 \times 12 = 19{,}354$.

34 $19{,}354$ (hours) $\div$ $1{,}981 = 9.8$ (hours per maintainer).

35 Traditional part-time guardsmen work approximately two days a month or 24 UTA days a year plus 14 days of annual training for a combined contribution of 38 days per year. We credit part-time personnel for helping to produce sorties 50 percent of the time, or 19 days per year (from discussions with 180 FW, Toledo). Multiplying 19 times 270 authorizations and dividing by the average availability factor for the active component (193 days), we get 27 FTEs.

36 Based on discussions with 180 FW, Toledo.

With 173 FTEs, the programmed flying hours per maintainer is 19.7 hours per person.[37]

To adjust the total number of FTE maintenance authorizations to the *effective* FTEs, the concept is to quantify the productivity lost because of the influence of the key factors. There are several kinds of productivity losses to consider.

First, some of the authorizations represent people who simply are not doing any actual maintenance at all, such as out-of-hide workers. Similarly, supervisors do not perform direct hands-on maintenance. A second category of loss is workers who are present and are doing maintenance work, but who are simply low in productivity. This category includes OJT trainers and trainees. A third category is an adjustment for availability. The notion here is that two equally productive individuals, that is, two maintainers who can do the same amount of work per unit time, will contribute to total unit productivity very differently if their available work hours per year are significantly different. The final category of adjustment is to estimate how many authorizations are effectively saved, or eliminated, in an ANG unit because of such management practices as cross-utilization and single-shift maintenance operations. These savings can then be projected onto an active component unit to estimate the number of manpower authorizations that are absorbed at the active unit because these techniques are *not* employed.

Regarding the personnel who are not doing hands-on maintenance, out-of-hide duties consume approximately 5 percent of the active component unit workforce. For the maintenance units at Shaw, this is approximately 99 people.[38] These 99 individuals are working, but they do not contribute to direct maintenance. So, 99 personnel are subtracted from Shaw's total authorizations to account for out-of-hide duties. The out-of-hide effect on the ANG is negligible; therefore, no adjustment is made to the Toledo ANG unit.

---

[37] 3,400 (hours) ÷ 173 = 19.7 (hours per maintainer).

[38] 1,981 (authorized maintainers) × 0.05 = 99 (personnel).

Shaw is authorized 400 three levels in its F-16 maintenance units.[39] Using the previously defined productivity estimates, a trainee contributes 40 percent productivity and a trainer contributes 85 percent productivity, compared with a fully trained maintainer. For Shaw, the active component OJT burden is approximately 300 people.[40] That is to say, approximately 300 personnel are absorbed in maintenance training. Therefore, 300 personnel are subtracted from Shaw's FTE personnel. For the ANG, OJT has little effect; therefore, no adjustment is made.

To account for maintainers in supervisory positions in the active component, all commissioned officers and E-7s through E-9s are subtracted from the total authorization. For Shaw, there are 204 supervisors.[41] For the ANG, a field team estimated the amount of time each senior enlisted leader or supervisor spent working on aircraft versus how much time was spent performing supervisory duties. The supervision burden for the Toledo unit is estimated to be 18 FTEs.[42]

Because the active component unit operates two-shift maintenance, an active unit has some extra authorizations that are not needed in an ANG unit. Some of these authorizations are driven by the inability to predict workload and some are needed because of minimum crew size requirements. The RAND team conducting this analysis developed a second-shift model[43] to help facilitate discussion on this issue. The model and its assumptions have been validated by the units visited in support of this analysis.[44]

---

[39] Based on the Shaw AFB unit manning document. Again, there may be more three levels assigned; however, this analysis looks only at authorized manpower. More assigned three levels could make the unit productivity even lower.

[40] Four hundred three levels times (1 – 0.4) trainee productivity equals 240 personnel plus 400 three levels times (1 – 0.85) trainer productivity equals 60 personnel, which is a total of 300 personnel.

[41] Based on the Shaw unit manning document.

[42] Based on discussions with 180 FW, Toledo, Ohio.

[43] See Chapter Three for more information about the second-shift model.

[44] Based on discussions with 180 FW, Toledo, and 20 FW, Shaw.

To open a minimum-sized second shift on an ANG F-16 base, 13 additional technicians would be needed to fully staff both shifts. Viewed in this manner, these positions represent overhead costs that would need to be absorbed regardless of how much work was done on the second shift. Thirteen authorizations represent approximately 7 percent of the 173 ANG FTEs. Applying this 7 percent factor to the active component unit at Shaw, the yield is approximately 138 personnel.[45] Using this logic, about 138 of the active component positions at Shaw are needed because Shaw runs a second shift. Thus, these 138 authorizations are subtracted from the Shaw total. Since the Toledo ANG unit does not run a second shift, no adjustment is made.

The ANG takes advantage of its highly experienced maintenance workforce through cross training and cross-utilization. Cross-utilization allows the ANG to employ a much smaller total workforce. Each individual shop can be smaller than would otherwise be required because each shop can count on outside help to handle its occasional peak workload. Field teams working with ANG personnel estimate that the Toledo workforce would need at least 12 more full-time authorizations (6.5 percent of its total) if this cross-utilization or sharing strategy were not employed.[46] While there is some difference in sortie generation workload for Shaw and Toledo,[47] there appears to be opportunity for the active unit to adopt this change. Applying this factor to the active unit at Shaw implies that 128 full-time authorizations could be saved if the active component units were structured more like the ANG workforce.[48] Since the ANG employs cross-utilization, no adjustment is made to Toledo.

Active component personnel have many responsibilities other than their direct maintenance tasks to accomplish. These other duties represent about one-half of a day per week that the active component

---

45 1,981 (personnel) × 0.07 = 138 (personnel).

46 Based on discussions with 180 FW, Toledo, and 20 FW, Shaw.

47 F-16s at Shaw AFB fly approximately 5.2 hours per week (19,354 PFH divided by 72 PAA divided by 52 weeks) compared with F-16s at the Toledo base, which fly approximately 4.4 hours per week (3,400 PFH divided by 15 PAA divided by 52 weeks).

48 1,981 (personnel) × 0.065 = 128 (personnel).

maintainer is not available to perform maintenance.[49] To calculate the active component availability, we compute the number of workdays per year (52 weeks times 5 days a week is 260) minus federal holidays (11 days) minus leave (30 days) minus the one-half of a day a week (26 days). This leaves a total of 193 workdays per year. Allocating one-half of a day per week for training (new, continuation, or another type of training) and additional duties is considered a cautious estimate.

A full-time ANG technician, in contrast, does not lose 26 days per year to outside duty requirements and training (maintenance skills and other training) because these are accomplished during traditional guard drill weekends. In addition, 15 days are not lost because of leave; the ANG technician uses leave to perform annual training. As a result, a full-time ANG technician is available to perform maintenance for 234 days per year. That is to say, a full-time ANG technician is 121 percent more available than an active duty maintainer on an annual basis.

It therefore follows that, if active duty personnel were as available as ANG technicians, an additional 193 positions would not be needed at Shaw. Note that this conclusion is in no way a criticism of Shaw or its management practices. This is simply an attempt to quantify the effect of the ANG environment, policies, and procedures on its manpower requirements and to compare them to active component experience. We summarize the quantitative effects of these factors in Table 2.4.

After considering the quantified effects of all the key factors as reductions to the FTE authorizations, the active component unit at Shaw is left with 919 FTEs,[50] which yields 21.1 programmed flying hours per FTE maintainer. The ANG unit at Toledo, in contrast, is left with 155 FTEs,[51] which yields 21.9 programmed flying hours per FTE maintainer. The difference in productivity between Shaw and Toledo on this basis is relatively minor. Thus, the key factors identified and

---

[49] Oliver, 2001; Albrecht, 1979; and Dahlman, Kerchner, and Thaler, 2002.

[50] 1,981 – 99 (out of hide) – 300 (OJT) – 204 (supervision) – 138 (shift) – 128 (cross-utilization) – 193 (availability) = 919 FTEs.

[51] 173 – 18 (supervision) = 155 FTEs.

Table 2.4
Key Factors Applied to Unit Productivity at Shaw AFB and Toledo AGS

	Shaw AFB	Toledo AGS
PFH	19,354	3,400
F/T authorizations	1,981	146
P/T authorizations	0	270
Total authorizations	1,981	416
FTE (FT + PT)	1,981	173
PFH/FTE	9.8	19.7
Out of hide	99	0
OJT requirement	300	0
Supervision	204	18
Shift or scheduling	138	0
Cross–utilization	128	0
Availability	193	0
Net effective	919	155
PFH/net effective	21.1	21.9

NOTE: These calculations could be considered cautious because other factors, such as split operations and assigned fill rates, are not considered in the calculations.

quantified in this analysis can be considered as accounting for the standards-based difference in productivity.

Figure 2.1 illustrates the relative importance of each of the key factors that influence maintenance productivity in the active component maintenance units at Shaw. Based on this analysis, a typical active component unit experiences approximately 47 percent of its maximum potential effectiveness per person assigned. In contrast, an ANG unit achieves about 90 percent effectiveness. This effectiveness differential

Figure 2.1
Relative Importance of Key Productivity Factors at an Active Component Maintenance Unit

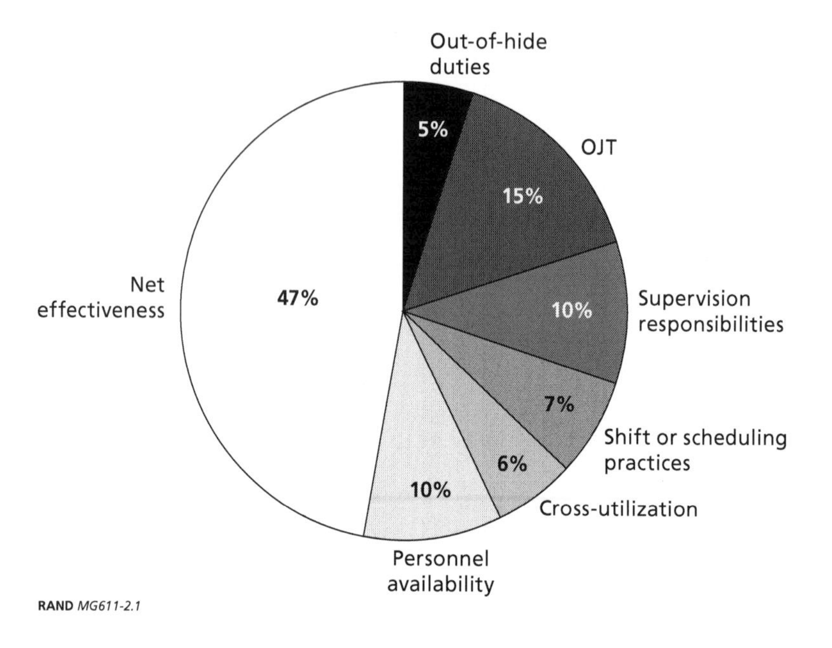

RAND *MG611-2.1*

can be directly attributed to differences between the ANG and active duty units in their training burdens, availability of manpower, experience levels, and related management practices.

CHAPTER THREE

Evaluating Options for Meeting Active Associate Maintenance Requirements

In the second part of the analysis, we evaluated active associate staffing requirements. Having identified the key factors and how they affect maintenance unit productivity, we used these insights to evaluate options for establishing an active associate unit for which the goal is to produce trained pilots in the most efficient manner possible. This evaluation includes unit-level daily flying programs and shift operations using simulation models and rules-based applications developed for this analysis.

RAND was asked to develop options for various maintenance workforce compositions based on the proposed flying hour programs and templates provided by the TFI offices. The TFI offices formulated flying hour programs for the active associate units that would satisfy pilot training requirements. Simply running an LCOM analysis for each proposed unit and its flying program may not be appropriate because the expected output of the active associate unit is trained pilots, not a wartime capability. In addition, the LCOM databases do not capture the differences in ANG operations and policies based on the relative efficiencies of ANG peacetime operations as previously described. Thus, the LCOM results should not be expected to accurately establish the peacetime, full-time positions that an ANG unit would require. Instead, we employ a micro-level scheduling analysis to determine second-shift requirements and, hence, maintenance authorizations needed to support a given flying scenario. These requirements are then adjusted based on differences in ANG and active duty productivity factors to define alternative staffing options.

ANG Full-Time Maintenance Manpower Model

The ANG uses a standard manpower model to determine its full-time authorizations. The model inputs include assigned aircraft and proposed flying hours and sorties. However, the primary driver is annual flying hours. Applying the ANG model, an ANG unit gains one full-time position for each increment of about 50 annual flying hours. The ANG manpower model implicitly assumes ANG efficiencies—for example, higher manpower availability, a single maintenance shift, and cross-utilization. While ANG full-time employees must be in a traditional ANG slot at the unit, there is no methodological connection between the ANG full-time model and the traditional workforce authorization generated by LCOM.

A review of the TFI active associate unit flying hour proposals (to satisfy the increased pilot training requirements) indicates that single-shift operations may not be adequate to accomplish the additional maintenance workload of the active associate unit.[1] A two-shift maintenance operation may be necessary. It was therefore necessary to develop a methodology to address second-shift authorizations.

RAND Methodology

Aircraft utilization and flight scheduling can affect the total maintenance workload, including the need for a second shift. Understanding these effects requires insight into the proposed flying hour program and aircraft flight scheduling, which are needed before active associate staffing requirements can be evaluated.

The ability of a unit to support its daily flying schedule depends on several components, the first of which is aircraft availability. Air-

[1] The TFI-proposed UTE rate is 18.4. See Appendix A for a list of TFI initiatives.

craft availability becomes a major concern as the unit flying hours increase. Assigned aircraft become unavailable to the daily flying program for several basic reasons, including major inspections (phase), programmed depot maintenance, scheduled depot upgrade maintenance, depot preparation or acceptance inspection after depot maintenance, scheduled time change or other major field-level maintenance, wash, or cannibalization (CANN jet).[2] These and other reasons prevent an aircraft from being available to fly. Thus, some number of a unit's total assigned aircraft generally will not be available to support the flying schedule. Moreover, some additional aircraft will inevitably become unavailable at unpredictable times because of failures or breaks that occur while performing the flying schedule. These aircraft must be repaired through unscheduled, on-equipment maintenance actions before they can be returned to the pool of available aircraft.

The effort required to support the daily flying schedule also depends on the number of required sorties per day and how they are grouped. The flying sequence (or the specific takeoff and landing sequence) and the number of times an aircraft will be flown per day are factors in the flying schedule that will significantly affect maintenance manpower requirements.

Given a pool of available aircraft and a specified daily flying schedule, modeling can simulate a unit's operation to determine whether a flying schedule can be supported with a single shift or whether the schedule is so demanding that two shifts will be needed to provide sufficient available aircraft.

Our model is based on the ability to meet the flying schedule—to ensure that adequate aircraft are always available to meet the required schedule. (The simulation model is described in more detail in Appendix C.) Simulation modeling of the daily flying schedule can be used to understand the interrelated effects of several factors, including flying days per week, number and timing of sorties each day, historical break rates, repair times, and the number of maintenance shifts. Inputs to the simulation model, as shown in Figure 3.1, include the number of

---

[2] A CANN jet is used to provide parts when the supply chain is unable to provide required parts as needed.

Figure 3.1
Overview of the RAND Model Used to Examine Scheduling Effects on Maintenance Requirements

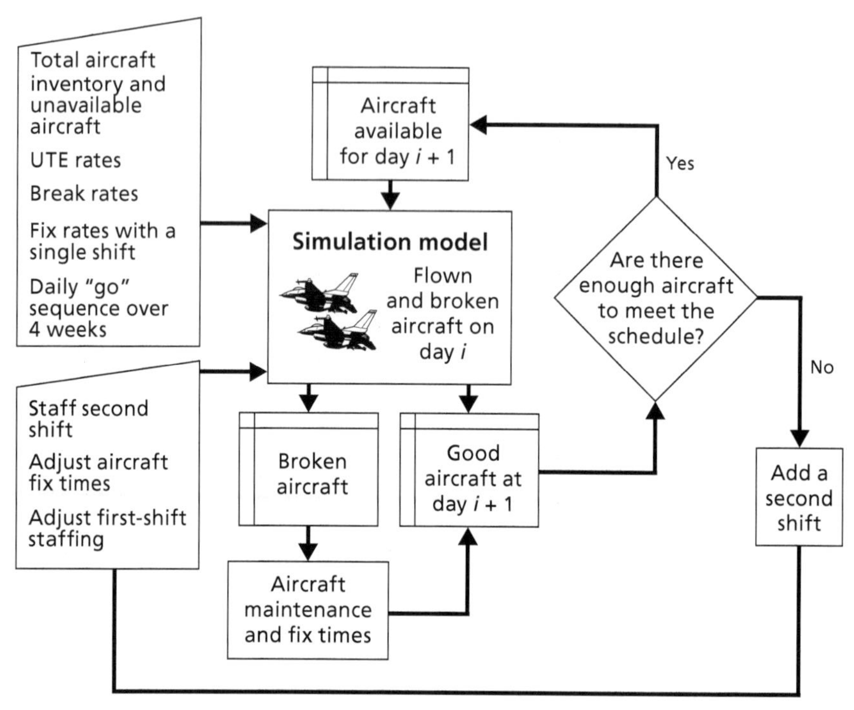

RAND *MG611-3.1*

aircraft assigned to the unit, the number of aircraft not routinely available for the flying schedule, the sortie utilization rate, the historical break rate, the historical 8- and 12-hour repair rate, and the daily sortie sequence (including number of days flown per week; the number of waves, or groups, of sorties; and the number of sorties in each wave). The simulation then steps through many simulated weeks of flying to study the supportability of the scenario with the available aircraft.

As aircraft return from their sorties, the model randomly breaks aircraft to match historical break rates and tracks the time needed to repair them. If the model calculates that the proposed schedule can be sustained with the required number of spare aircraft using the available aircraft, then the sequence continues. If it cannot be sustained, the model tracks the number of lost sorties and the number of days without sufficient launch spares. When the model deems the proposed schedule unsustainable, a second shift can be added, and repair times are adjusted to account for the second shift.

Model Application to the TFI Template

The active associate unit initiative proposed by TFI involves moving active component pilots to an ANG base for pilot training. Supporting the additional pilots would require additional training sorties, increasing the aircraft UTE rate at the designated ANG unit.

BRAC has called for moving aircraft from closing locations to increase the PAA at other locations, some at ANG units. However, the receiving units will not gain additional pilots or maintainers from the BRAC-directed moves. Therefore, ANG units receiving additional aircraft anticipate actually flying a lower UTE rate (the same number of hours but spread across more aircraft), at least initially.

The evaluation of staffing requirements for an active associate unit should consider the additional PAA added by BRAC, the higher UTE rate implied by TFI pilot absorption goals, and the potential consequences of adding a second maintenance shift (see Tables 3.1 and 3.2). Based on the ANG full-time manning model, maintaining the current ANG UTE rate (15) with the BRAC-driven increase in PAA (from 15 PAA to 18 PAA) would yield 4,212 annual programmed flying hours (an increase of 702 from the current 3,510). To maintain 4,212 programmed flying hours, an additional 14 full-time technicians would be needed to accomplish the increased maintenance workload, as established by the ANG Full-Time Maintenance Manpower Model. When the TFI-proposed UTE rate of 18.4 is examined (an increase of 3.4), the programmed flying hours increase to 5,167. The corresponding maintenance manpower would increase by the original 14 (driven by BRAC) plus 18 more (driven by the increase in the UTE rate) for an overall increase of 32 positions or a total workforce of 178 full-time technicians.

Table 3.1
Increased PAA and UTE Rate Effects on the Active Associate F-16 Unit—Unit Metrics

		TFI Scenarios	
	Current	One Shift	Two Shifts
PAA	15	18	18
UTE rate	15	18.4	18.4
ASD	1.3	1.3	1.3
PFH	3,510	5,167	5,167

NOTE: The "current" column does not represent any specific unit. Rather, it is a generalized view of what an ANG unit with 15 PAA is accomplishing today. The manpower breakout, however, is closely modeled after the 180 Fighter Wing (FW), Toledo AGS.

Table 3.2
Increased PAA and UTE Rate Effects on the Active Associate F-16 Unit—Manpower Authorizations

		TFI Scenarios	
	Current	One Shift	Two Shifts
Staff	19	22	26
AMX	56	66	68
Maintenance Group leadership	4	4	5
EMS	26	33	33
CMS	41	53	59
TOTAL	146	178	191
TFI UTE rate Δ		32	32
2nd Shift Δ			13
Total Increment		32	45

NOTE: The "current" column does not represent any specific unit. Rather, it is a generalized view of what an ANG unit with 15 PAA is accomplishing today. The manpower breakout, however, is closely modeled after the 180 Fighter Wing (FW), Toledo AGS.

These manpower increases implicitly assume standard ANG maintenance practices, including a single shift. As demonstrated by the simulation model, this higher level of flying activity may necessitate a second shift.

The ANG currently is able to have sufficient aircraft available so that the flying schedule can be met on a daily basis with a single maintenance shift. If an ANG unit were to move from its relatively low current sortie rate to the higher utilization rates implied by the TFI fighter pilot templates, the unit may no longer be able to support peacetime training sortie generation with single-shift maintenance. Single-shift versus two-shift operation is a function of how stressful the flying schedule is, that is, how many aircraft must be available at a given point in time and how willing the unit may be to trade off the mission-capable aircraft rate in favor of a single-shift maintenance operation.

As discussed previously, the supervision and other overhead costs associated with opening a second shift at an ANG unit are approximately 13 positions (see Chapter Two). If the second shift were established, some of the workload (and hence some of the technicians) from the first shift would transfer to the second shift. The total increase to move from a 15 PAA ANG squadron flying a 15 UTE rate to an 18 PAA ANG unit flying an 18.4 UTE rate with a second shift would be approximately 45 additional full-time technicians. This increase would meet requirements only for the peacetime training of pilots. It does not account for any changes that may be required for the unit's wartime capability.

There are several ways in which the Air Force could meet the increased workforce requirements at an active associate unit. For example, the number of full-time ANG technicians could be increased to meet the new requirement. That is, the ANG unit could simply convert some of its traditional slots to full-time technician slots and hire personnel to fill these authorizations. Alternatively, the Air Force could establish a cadre of active duty maintainers at the guard unit to provide the needed maintenance capability. In addition, the Air Force might choose to make some or all of the maintenance cadre trainee-level personnel. In this way, the active associate unit could be seen as providing training not only to active duty pilots, but also to active duty maintenance personnel.

The total requirement to satisfy an 18.4 UTE rate at an 18 PAA unit while operating a second maintenance shift would be 191 full-time technicians (see Tables 3.1 and 3.2). This notional unit currently has 146 full-time personnel authorized. The TFI requirement would add 45 full-time ANG technicians or, because of the lower availability of active duty personnel, an additional 54 fully qualified active component maintainers.[3] RAND field teams working with ANG shop chiefs estimate that an ANG maintenance unit could absorb a total of 22 active duty junior maintenance trainees without undue disruption. Using the previously identified productivity factors associated with OJT, these 22 maintenance trainees would require an additional four FTE personnel to accommodate the trainer workload requirement, and thus an additional 36 full-time ANG technicians would also be needed to bring the unit to full strength (see Tables 3.3 and 3.4).

OJT for maintenance personnel is a burden to the active component that the ANG full-time maintenance technicians do not face. For the purpose of this analysis, all E-3s and below are considered trainees. Approximately 20 percent of the maintainers at Shaw AFB, or approximately 400 personnel, are considered trainees. The productivity of the 400 maintenance trainees (at 40 percent productive) equates to 160 fully productive people.[4] In addition, 400 of the E-5s and E-6s at Shaw must act as maintenance trainers. As discussed in Chapter Two, trainer productivity is estimated at 85 percent. However, at Shaw AFB, there are more than 400 E-5s and E-6s. Thus, the overall productivity for E-5s and E-6s is 92 percent.[5] At Shaw, the maintenance OJT burden alone—with 20 percent trainees—will reduce the number of effective FTEs from the 1,928 enlisted personnel authorized to the equivalent

---

[3] 45 × 1.21 = 54 (active component maintainers). However, these are fully qualified active maintainers, the equivalent of a seven-level maintainer.

[4] See Chapter Two for discussion of trainee and trainer productivity.

[5] Four hundred E-5s and E-6s are 85 percent productive. The rest are 100 percent productive, yielding an overall productivity of 92 percent.

Table 3.3
Options for Fulfilling TFI Staffing Scenario Requirements—Unit Metrics

TFI Staffing Scenario Variables	Quantity
Existing full-time ANG staff	146
Increment with single shift	32
Increment with two shifts	45
Trainee productivity	40%
Trainer productivity	85%
ANG substitution ratio	1.21

Table 3.4
Options for Fulfilling TFI Staffing Scenario Requirements—Manpower Authorizations

	Now	ANG Maintainers	Active Component Maintainers	Trainers	Active Component OJT	Total
With one shift						
All full-time ANG maintainers	146	32				178
All fully qualified active component maintainers	146		39			185
Maximum active component OJT absorption	146	23		4	22	195
With two shifts						
All full-time ANG maintainers	146	45				191
Maximum active component participation	146		54			200
Maximum active component OJT absorption	146	36		4	22	208

of 1,628 fully trained (productive) maintenance personnel (see Table 3.5).

If the maximum potential capability of the ANG to absorb F-16 maintenance trainees were exercised, this would amount to about 22 trainees at each of 19 ANG F-16 units. If these trainee authorizations were therefore reduced proportionately at all ACC F-16 units, the reduction at Shaw would be about 80 maintenance trainee positions. This change would represent a sharing of Shaw's OJT burden with the ANG. As is shown in Table 3.5, this sharing would reduce the total authorizations at Shaw by 80 positions, that is, from 1,928 to 1,848. Because of the lower productivity of trainees and their trainers, however, the effective FTE total is reduced by only 20, that is, from 1,628 to 1,608.

While transferring active component maintenance trainees to the ANG to help satisfy the increased requirements driven by the associ-

Table 3.5
Quantifying the On-the-Job Maintenance Training Burden at Shaw AFB

	All OJT at Shaw AFB			Some OJT at Associate Unit		
Grade	No. of Slots	Effectiveness	Net	No. of Slots	Effectiveness	Net
E-9	13	100%	13	13	100%	13
E-8	34	100%	34	34	100%	34
E-7	129	100%	129	129	100%	129
E-6 (trainer)	246	92%	226	246	93%	230
E-5 (trainer)	488	92%	448	488	93%	456
E-4	618	100%	618	618	100%	618
E-3 (trainee)	400	40%	160	320	40%	128
Total	1,928		1,628	1,848		1,608
Total percentage			84.4%			87.0%

NOTE: The assumptions are as follows: Trainee productivity is 40 percent; there is 1.0 trainee per trainer; trainer productivity is 85 percent.

ate unit may seem appealing from the active unit's point of view, the Air Force must consider the full impact of losing these trainees. By removing 80 maintenance trainees from the unit at Shaw, the equivalent of 20 fully effective personnel is lost. This amount may seem like a small reduction in capability. As we have shown in the previous chapter, however, the active component units may be at or near full production capability now. The loss of 80 personnel, even though their production is low, may be very difficult to absorb because not every task in a maintenance unit requires a fully qualified individual. However, with some of the maintenance training burden relieved from the active component five and seven levels, some of the productivity will be recaptured. Instead of being only 85 percent productive while acting as trainers, these experienced maintainers will be able to be fully productive, negating some of the loss of the three levels.

Personnel flow constraints should also be considered[6] when deciding how many maintenance trainees could be transferred to the ANG. The personnel flow constraint identifies the necessary ratio of active component personnel to reserve component personnel that is sufficient to provide a critical flow of human capital from the active component to the reserves. The active component should retain enough positions in the maintenance career field to train future recruits for the ANG. Although this limitation does not appear to be severe, it should be considered before establishing active associate maintenance units.

The simplest way to satisfy the additional maintenance requirements of an active associate unit may be to hire more full-time ANG personnel to perform the additional maintenance. This method does not increase either active component or reserve component total authorizations (which would not be allowed by PBD 720) since the ANG could convert traditional slots to full-time slots. It does not reduce active component productivity since no active component personnel are moved to an associate unit and it would preserve the personnel flow for the ANG. Of course, this strategy would require a budget increase for the reserve component units. However, sharing some of the main-

---

6 Robbert et al., 1999.

tenance OJT burden between the active and reserve components may also be worthy of consideration.

CHAPTER FOUR

Summary of Findings

The ANG is consistently able to generate more peacetime flying hours per full-time-equivalent maintainer than is the active component—in the case of the F-16, more than twice as many. There are several factors that contribute to this difference in productivity. First, ANG units possess a highly experienced workforce, not only in maintenance but in many career fields. Historically, ANG maintainers remain in the same location much longer than their active component counterparts, and this stability allows them to develop deep and broad job knowledge. Because of their extensive knowledge, many ANG personnel are cross trained and cross utilized. ANG personnel are able to work the flight line as well as being certified to work in a back shop.

Second, the traditional ANG unit recruits experienced personnel from the active component or the civilian sector. Again, this point applies not only to maintenance but also to many other career fields. ANG full-time positions are filled with highly qualified, fully trained individuals. Thus, the unit is able to spend most of its time on direct production tasks and very little time performing initial or upgrade training. When continuation or other training is required, the ANG workforce can use drill weekends to complete the training. Also, a highly experienced workforce inherently requires less supervision.

The active component, on the other hand, has a large number of inexperienced maintainers who require hands-on training and supervision. The active component does not have extra weekends (drill weekends) on which to complete skill development, refinement, or other training. All training—maintenance and otherwise—occurs during the normal workday. The active component member also has other military duties that reduce the time available to perform hands-on maintenance.

Finally, the typical active component unit operates two shifts, which can require a minimum crew size on each shift. While two shifts can make a unit very effective, it is not as manpower efficient as a single-shift operation. Most ANG units, on the other hand, provide only a single shift to support peacetime flying training.

The methodology developed in this research can be used to quantify and compare the key factors that allow the ANG to generate peacetime training sorties with a fairly small full-time workforce. By applying these insights to proposed future operations and the proposed TFI associate unit initiatives, staffing alternatives can be suggested that demonstrate how various types of personnel can influence the size and productivity of a proposed unit. These insights can be used to develop reasonable maintenance options for supporting additional flying requirements, such as those associated with TFI initiatives.

It is clear that OJT is a burden to the active component that the ANG full-time technicians do not face. Maintenance options for supporting TFI associate unit initiatives can also be developed to reduce the OJT burden on the active component by sharing this burden with the ANG.

The methodological approach provided in this monograph can be used as a framework to analyze the best practices for staffing an active associate unit. The general findings on maintenance productivity differences between ANG and active duty units, and the root causes of these differences, may also provide useful insights into the more general issue of increasing the productivity of Air Force operations.

APPENDIX A

Total Force Integration Initiatives

The Future Total Force concept, now called Total Force Integration (TFI), was developed by the U.S. Air Force to leverage capabilities in each component of the Air Force—the active component and the reserve component. Listed below are TFI initiatives currently being considered. The initiatives are listed by state.[1]

Alaska

- Create a classic associate unit with AFRC on two 18 PAA F-22A squadrons at Elmendorf AFB.
- Create a Classic Associate with AK ANG on 8 C-17s at Elmendorf AFB.

Arizona

- Create an active associate with ANG on 8 KC-135s at Phoenix Sky Harbor International Airport (IAP).
- Reallocate manpower and equipment from KC-135 unit to stand up an operational ANG Predator squadron at a location to be determined.

---

[1] The initiatives are from U.S. Air Force, *Watch List,* U.S. Air Force, Directorate of Total Force Integration, Mission Development Division, May 4, 2006.

- Investigate a classic associate unit with AFRC for A-10 FTU at Davis-Monthan AFB.
- Continue the Joint Warfighter Space Mission in Arizona if manpower is available—if no manpower is available, consider other state options in which manpower is available.

California

- Establish a formal training unit for Predator launch and recovery. Determine a long-term Predator beddown solution among different possible courses of action at various locations, including Edwards AFB.
- Establish an ANG unit equipped operational Predator squadron at March Air Reserve Base (ARB). Establish a maintenance field training detachment through close coordination with Air Education and Training Command (AETC).
- Create a classic associate unit between active and reserve components to augment Global Hawk and other operations at Beale AFB.
- Expand classic associate with AFRC and create a classic associate with ANG for a space group or wing at Vandenberg AFB—supporting the Joint Space Operations Center (JSPOC).
- Create a classic associate with ANG for a space group or wing at Vandenberg AFB in support of the JSPOC at Vandenberg AFB.

Colorado

- Establish an ANG and investigate a reserve associate unit for the Rapid Attack, Identification, Detection, and Reporting System (RAIDRS).
- Investigate a reserve associate unit for RAIDRS at Peterson AFB.
- Establish a classic associate unit with AFRC for a space-based infrared system mission control station (SBIRS MCS).
- Establish a C-130 active associate unit at Peterson AFB.

Connecticut

- Create an active associate unit with ANG at Bradley IAP for a BRAC-directed TF-34 (A-10) engine CIRF.
- Establish an ANG Joint Cargo Aircraft (JCA) unit at Bradley IAP.

District Of Columbia

- Establish a Maryland ANG–D.C. ANG fighter unit initiative.
- Investigate an active/ANG/AFRC association at Andrews AFB.

Florida

- Fully integrate Total Force F-22A FTU (ANG/AFRC/active) at Tyndall AFB.
- Create a BRAC-directed classic associate unit with AFRC on 16 KC-135s at MacDill AFB.
- Establish a BRAC-directed U.S. Air Force F-15 avionics CIRF at Tyndall AFB.
- Create War Fighting Headquarters (WFHQ) manpower augmentation with AFRC in support of Air Force Special Operations Command (AFSOC) at Hurlburt Field.
- Enhance ANG space mission at Patrick AFB (Range Operations).
- Establish an active/AFRC classic associate with ACC units at Eglin AFB.
- Establish an AFSOC WFHQ at Hurlburt Field.

Georgia

- Establish a BRAC-directed U.S. Air Force TF-34 (A-10) engine CIRF at Moody AFB.
- Investigate an AFRC classic associate in A-10s at Moody AFB.

Hawaii

- Create a classic associate unit with ANG on eight C-17s at Hickam AFB.
- Establish a BRAC-directed active associate unit with ANG on 12 KC-135s at Hickam AFB.
- Create ANG and AFRC WFHQ augmentation with U.S. Air Force, Pacific Command.

Illinois

- Establish a BRAC-directed ANG F-110 (F-16) engine CIRF at Abraham Lincoln Capital Airport, Springfield.
- Provide ANG manpower to support a BRAC-directed logistics support center at Scott AFB.
- Establish a C-40C integration/active association at Scott AFB.

Indiana

- Establish a new ANG air support operations squadron (ASOS) at Hulman Regional airport.
- Create an ANG distributed ground station (DGS) (location to be determined).

Louisiana

- Create an active associate unit at New Orleans Naval Air Station (NAS) Joint Reserve Base (JRB) with ANG to form a BRAC-directed F-15 engine CIRF.

Maryland

- Establish a Maryland ANG–D.C. ANG fighter unit initiative.

- Investigate the ANG JCA unit at Martin State APT.
- (See District of Columbia for active/ANG/AFRC associate unit at Andrews AFB).

Massachusetts

- Create an ANG DGS at Hanscom AFB.

Missouri

- Create a classic associate unit with ANG on 16 B-2s at Whiteman AFB (pending resolution of a personnel reliability program issue for the ANG and AFRC).

Nevada

- Continue the ANG/AFRC association throughout missions at the U.S. Air Force Warfare Center, Nellis AFB.

New York

- Create a BRAC-directed Air Reserve Component (ARC) Associate Unit with AFRC as lead on eight C-130s at Niagara IAP/Air Reserve Station (ARS). A minimum of four additional C-130 aircraft are required, with eight additional C-130s desired to create a viable associate.
- Activate an MQ-9 squadron within the New York ANG. It is understood that final funding decisions may be delayed until the final production decision on the MQ-9 is made in FY08.

North Carolina

- Create a BRAC-directed active associate unit with AFRC on 16 C-130s at Pope AFB.
- Create a BRAC-directed active associate with AFRC on 16 KC-135s at Seymour Johnson AFB.
- Investigate an AFRC classic associate in F-15Es at Seymour Johnson AFB.
- Establish an F-100 engine BRAC-directed CIRF at Seymour Johnson AFB.

North Dakota

- Establish an ANG JCA unit at Hector IAP.
- Continue the programmed plan for ANG security forces unit at Minot AFB.
- Establish Predator/family of unmanned aerial system (UAS) missions initiative.
- Develop a beddown and operational BRAC-directed plan to support Global Hawk missions.
- Establish an ANG operational Predator squadron at Hector IAP.

Ohio

- Create a foreign military sales F-16 training mission at Springfield-Beckley MAP, leveraging the current ANG FTU experience and infrastructure.
- Create a measurement and signature intelligence classic associate unit with ANG supporting the National Air and Space Intelligence Center (NASIC) at Wright-Patterson AFB.
- Investigate the ANG JCA unit at Mansfield Lahm airport.

Oklahoma

- Create a BRAC-directed ARC associate unit with AFRC as lead on 12 KC-135s at Tinker AFB.

Oregon

- Establish a combat support wing (manpower neutral).

Pennsylvania

- Establish a new ANG ASOS unit.

South Carolina

- Create a BRAC-directed electronic countermeasures (ECM) pod CIRF at Shaw AFB.
- Create a classic AFRC association with 20 FW F-16s at Shaw AFB. AFRC 307FS, Det1 Shaw currently provides a small number of pilots and maintenance personnel to one F-16 squadron in 20 FW through the AFRC Fighter Reserve Associate Program (FRAP).

Tennessee

- Establish an ANG intelligence production squadron in support of ACC at Nashville Metro airport.
- Enhance an ANG command and control squadron in support of space operations at McGhee-Tyson airport.
- Investigate an ANG JCA unit at Nashville Metro Airport (MAP).

Texas

- Activate an AFRC C-5 FTU at Lackland AFB.
- Investigate a new ANG ASOS unit.
- Stand up an ANG operational Predator squadron at Ellington Field.

Utah

- Continue the CSAF initiative for a classic associate unit with AFRC on 72 F-16s at Hill AFB. The CSAF Total Force test initiative is on track.
- Establish an active duty unit for a BRAC-directed targeting pod and F-110 engine CIRFs at Hill AFB.

Vermont

- Continue the CSAF-directed community basing initiative. The Total Force Initiative is on track to evaluate assignment of active duty personnel to ARC unit not colocated on or near an active duty military facility.

Virginia

- The 192 FW will associate with the 1 FW in its F-22A mission and any additional missions as agreed upon by ANG, ACC, 1 FW and the adjutant general of Virginia.
- Continue the AFRC association with 1 FW in the remaining F-15C unit. The AFRC 307 FS Langley flight provides a small number of pilots and maintenance personnel to one F-15 squadron in 1 FW through the AFRC FRAP.

Washington

- Create a BRAC-directed classic associate unit with ANG at Fairchild AFB.
- Continue the combat support wing initiative.

Wyoming

- Create a BRAC-directed active associate with ANG on 12 C-130s at Cheyenne MAP.

Puerto Rico

- Establish an ANG JCA unit.

APPENDIX B

F-16 Unit Productivity Comparisons

This appendix presents productivity examples from three F-16 bases—Shaw AFB, McEntire AGS, and Toledo AGS. As Table B.1 shows, there are significant differences in productivity between ANG and active component units in producing peacetime training sorties. An ANG F-16 unit is able to generate more than twice as many peacetime training flying hours per FTE maintenance authorization as a comparable active component unit.

Table B.1
Comparison of Active Component and ANG F-16 Programmed Flying Hours per FTE Authorizations

	Combat Coded F-16 Units		
	Shaw AFB	McEntire AGS	Toledo AGS
PFH	19,354	3,996	3,400
FT authorizations	1,981	163	146
PT authorizations	0	274	270
Total authorizations	1,981	437	416
FTE	1,981	190	173
PFH/FTE	9.8	21.0	19.7

APPENDIX C

RAND Scheduling Model: Simulation of the Sortie Generation Process

The current maintenance practice in ANG F-16 flying units is to generate normal peacetime training sorties with a single-shift maintenance operation. This practice means that aircraft maintenance is routinely performed during an eight-hour duty day, five days per week. While the ANG has been successful with this strategy, the approach is feasible only when sufficient aircraft can be made available so that the flying schedule can be met on a daily basis. This feasibility, in turn, is a function of how stressful the flying schedule is—that is, how many aircraft must be available at a given point in time and how willing the unit may be to trade off the mission-capable aircraft rate in favor of a single-shift maintenance operation. If an ANG unit were to move from its relatively low current sortie rate to the higher utilization rates implied by the TFI fighter pilot templates, the unit might no longer be able to support peacetime training with single-shift maintenance. To explore this idea, we developed a simulation model that can compare the results of single-shift and two-shift maintenance strategies. We can use this model to estimate whether a given sortie generation requirement (monthly flying schedule) can be satisfactorily achieved with a single-shift maintenance operation or whether a two-shift maintenance operation would be required.

The simulation is a traditional Monte Carlo discrete-event simulation that operates at a fairly high level of aggregation. It is important to recognize that this tool, unlike the more familiar and much more complex LCOM simulation, is not intended to size the maintenance manpower or other resources required to support a given flying program. Rather, this simulation tool focuses simply and directly on the following question: given a daily flying schedule, the break rates per sortie for the aircraft, and the distribution of repair times associated with these break rates, can sufficient aircraft be kept serviceable to sustain the flying schedule? Both the break rates and the repair times are assumed to be exogenous variables, which is the equivalent of saying that the analysis assumes that sufficient manpower and other resources will be made available to maintain these repair times.

The simulation model is embedded in an Excel spreadsheet. In concept, the model resembles a simple Materials Requirement Planning (MRP) tableau, in which the daily sortie requirement takes the place of the period demand and the expected available aircraft takes the place of the inventory. The simulation then removes aircraft from the pool of available aircraft as they experience random failures after sorties, and serviceable aircraft are returned to the pool of available aircraft after a randomly assigned repair time interval has passed. This process is repeated over many weeks of simulated flying and maintenance activity, and statistics are accumulated that show the number of sorties that were lost because of unavailability of aircraft and the number of instances in which an insufficient number of spare aircraft were available to support the daily flying schedule. In this way, the simulation can suggest whether the planned flying schedule could be supported with single-shift maintenance or whether two shifts might be necessary.

To illustrate the process, suppose a unit has 18 PAA and an additional 2 spare aircraft. If the unit maintains a UTE rate of 18.4 sorties/PAA/month, it will need to generate 3,974 sorties per year. With five flying days per week, the unit needs to generate about 16 sorties per day. An active duty unit would typically schedule a first wave of 12 sorties, typically in the morning. About four hours later, after the first wave had been recovered and serviced, a second wave of 4 sorties would be launched. To ensure that each wave could be launched, the unit would also require that at least 2 spare aircraft be ready to substitute for aircraft that might ground abort on launch. Thus, this flying schedule would require that 14 aircraft (12 plus 2 spares) be available each flying day to support the first wave requirement.

While the unit has a total of 20 aircraft to work with, about 5 aircraft will typically not be available to support the daily flying schedule. We would expect 1 aircraft to be in programmed depot maintenance, 1 to be undergoing its "phase inspection," 1 to be on the wash rack, 1 to be used as a maintenance trainer, and 1 to be the "CANN bird," that is, to have been cannibalized for spare parts. Thus, the unit must be able to consistently generate 14 of the remaining 15 aircraft to be able to fully support this flying schedule.

The break rate experienced by ACC for its F-16 fleet averages about 12 percent. This break rate implies that a unit will generate 1 failed aircraft (that is, 1 aircraft that requires unscheduled on-equipment remedial maintenance to become flyable) per 8 sorties. Typical fix rates experienced by ACC F-16 show that about 50 percent of these maintenance actions can be completed within 8 hours. Another 35 percent of these incidents can be completed within 12 hours, and the remaining 15 percent of the maintenance actions can be resolved within 48 hours. These empirical fix rates have been experienced by ACC F-16 units, which typically operate two full shifts of aircraft maintenance. We can, therefore, interpret these fix rates as estimates of the elapsed time or clock time required to complete the various maintenance tasks associated with on-equipment remedial maintenance. In other words, we consider that 50 percent of the maintenance actions are such that they can be completed in 8 clock hours, given that a shift is available to perform the work. Similarly, the 35 percent of incidents that an ACC unit can resolve in 12 hours can be assumed to require about 12 hours of shift labor to be completed. Note that with a routine single-shift maintenance operation, these 12-hour repair incidents would take two days to resolve, and so the availability of the aircraft would be reduced as a result. For those 15 percent of repair incidents that require 48 clock hours to complete, we recognize that over a 48-hour or two-day interval, a two-shift maintenance operation would provide a total of 32 hours of shift labor (two days times two shifts times 8 hours per shift). We, therefore, estimate that this class of repair action would require

32-shift hours to complete. With a single-shift maintenance operation, these aircraft would be unavailable for four days.

Within the simulation, we can use the break rate to randomly remove failed aircraft from the pool of available aircraft based on daily flying activity. On each flying day, we generate the sortie requirement, or as many sorties as can be provided by the aircraft available at that time. We randomly generate a number of failed aircraft using a binomial probability distribution. We can also use the fix rate data to deduce when these aircraft would be returned to the pool as flyable. Each failed aircraft is randomly assigned a required maintenance action type of 8, 12, or 32 hours, based on the proportions implied by the ACC break rate data, using a multinomial distribution.

We can now use these shift labor times to infer when a given aircraft will become available based on the maintenance schedule, that is, the number of maintenance shifts that are used. Peacetime flying operations are generally conducted in two waves. The first wave, normally in the morning, launches and recovers a set of aircraft. The preflight, takeoff, training sortie, and recovery actions generally take about four clock hours. Given an 0800 start of day, the first wave generates failed aircraft at roughly noon. A second wave of sorties is generated in the afternoon. This wave also takes about four hours to complete. Thus, aircraft failures generated by the second wave can enter the remedial maintenance process at the beginning of the second shift if there is one. If not, these aircraft will not be worked on until the next morning.

A typical ACC unit operates with two-shift maintenance, which means that an aircraft that flew in the first wave on, for example, Monday and failed would be available to receive up to 12 hours of maintenance on Monday. As a result, an aircraft that was randomly assigned an 8-hour repair time would not be available for the second wave on Monday, but it would be available for the first wave on Tuesday. The same would be true for an aircraft assigned a 12-hour repair time. An aircraft that is assigned a 32-hour repair time would be available for the first wave on Wednesday, that is, 48 clock hours later.

If an aircraft that flew in the second wave on Monday required maintenance, it would be available to receive up to 8 hours of maintenance on Monday before the second shift ended. As a result, aircraft assigned an 8-hour maintenance interval would be available for the first wave on Tuesday. An aircraft that failed in the second wave on Monday and that was assigned a 12-hour maintenance interval would become available for the second wave on Tuesday. An aircraft that failed in the second wave on Monday and that was assigned a 32-hour maintenance interval would become available for the first wave on Wednesday.

A typical ANG F-16 unit operates with single-shift maintenance, which means that an aircraft that flew in the first wave on, for example, Monday and failed would be available to receive only 4 hours of maintenance on Monday. As a result, an aircraft that was randomly assigned an 8-hour repair time would not be available until the second wave on Tuesday. A failure from the first wave on Monday that required 12 shift hours would not be available until the first wave on Wednesday. An aircraft failure from the first wave on Monday that is assigned a 32-hour repair time would not be available until four days later.

If an aircraft that flew in the second wave on Monday required maintenance, given a single maintenance shift, no shift labor would be scheduled until Tuesday morning. As a result, aircraft assigned an 8-hour maintenance interval would not be available until the first wave on Wednesday. An aircraft that failed in the second wave on Monday and that was assigned a 12-hour maintenance interval would become available for the second wave on Wednesday. An aircraft that failed in the second wave on Monday and that was assigned a 32-hour maintenance interval would not become available until five days later.

Given this understanding of the aircraft unavailability implied by flying schedules, break rates, fix rates, and maintenance schedules, we can construct a simple Monte Carlo simulation of unit flying and maintenance activity patterned after an MRP tableau. In this tabular array, each column of data represents a single day. Each row in the table represents an entity in the simulation. To illustrate the process, consider the example in Table C.1.

In this example, we have 15 available aircraft. We would read the table as follows. On Monday, we begin in row 1 with all 15 aircraft available. Since this is sufficient, in row 2, we launch a full first wave of 12 sorties. In row 3, we randomly generate two failures as a result

Table C.1
RAND Model Sample Unit Flying and Maintenance Activity

Row	One-Shift Scenario	Before Start of Mon Shift	Mon	Tues	Wed	Thurs	Fri	Sat	Sun
1	Available aircraft		15	13	15	15	14	10	15
2	First wave		12	12	12	12	12	0	0
3	First wave failures		2	0	0	0	3	0	0
4	8-hour repair		1	0	0	0	2	0	0
5	12-hour repair		1	0	0	0	1	0	0
6	48-hour repair		0	0	0	0	0	0	0
7	Available aircraft		13	14	15	15	11	13	15
8	Second wave		4	4	4	4	4	0	0
9	Second wave failures		0	0	0	1	1	0	0
10	8-hour repair		0	0	0	0	1	0	0
11	12-hour repair		0	0	0	1	0	0	0
12	48-hour repair		0	0	0	0	0	0	0
13	Total failed aircraft		2	0	0	1	4	0	0
14	Remaining available aircraft	15	13	14	15	14	10	13	15
	Total sortie production		16	16	16	16	16	0	0

NOTES: There are 15 available aircraft, a first wave requirement for 12 sorties, and a second wave requirement for 4 sorties—from Monday through Friday.

of this flying activity. In rows 4, 5, and 6, we randomly assign repair times to these two maintenance requirements; in this case, one is an 8-hour job and one is a 12-hour job. In row 7, we see that only 13 aircraft remain available; however, this is more than enough to launch a full second wave of 4 sorties (row 8). In row 9, we randomly assign a number of failures; in this case, none. Hence, no repair times are assigned in rows 10, 11, and 12. As a result of Monday's activity, 2 aircraft failures were generated (row 13), 13 aircraft remain available (row 14), and daily sortie production was 16 sorties, as scheduled.

This process is repeated, day by day (column by column) for as many simulated days as is desired. The only difference is that in subsequent days, repaired aircraft are added back to the available pool using the time delay logic outlined above. For example, there are 14 available aircraft for the second wave on Tuesday (row 7), 13 from row 1 (aircraft that rolled over from the end of the day on Monday (row 14), plus the eight-hour repair action from Monday, row 4, which is now completed. In this way, each completed maintenance action results in a time-phased return of an aircraft to the available pool.

In this way, we can simulate the sortie generation and repair processes over time. We gather output statistics on the number of sorties generated, as well as the number of situations in which the requirement of two additional available spare aircraft is not achieved. Table C.2 illustrates the results of a 32-week simulation of the flying schedule described above.

With a 12-sortie first wave and a 4-sortie second wave, five days per week, the schedule calls for 80 sorties per week or a total of 2,560 sorties over 32 weeks. With a single shift, the simulation indicates that only 2,497 sorties would be launched. A two-shift operation performs much better, generating 2,559 sorties. We also accumulate statistics on the number of spare aircraft short per day; that is, wave launches in which there are fewer than two flyable spare aircraft available. Here we see that, with two maintenance shifts, we are occasionally missing a spare aircraft. In this simulation, we miss 27 spares, or a little less than one per week. With single-shift maintenance, the results are much worse. The unit is short 207 spares over the 32-week interval. Note also how the problem builds over the week. The unit is usually in good shape on Mondays, but, as the failures accumulate, by Thursday and Friday, the unit is routinely short one or both of the required spares. This situation would probably not be tolerable to the unit. In addition, the reason the unit can recover its position on Monday is that we assume that the single-shift maintenance will proceed as overtime on Saturday and Sunday on an as-required basis, that is, if the unit needs these aircraft for Monday flying. Occasional overtime is toler-

Table C.2
Simulation of a 32-Week Flying Schedule

Sample One-Week Schedule

Schedule	Mon	Tues	Wed	Thurs	Fri	Sat	Sun	Total
First wave	12	12	12	12	12	0	0	60
Second wave	4	4	4	4	4	0	0	20

32-Week Schedule

Schedule Target	2,560
Sorties, one shift	2,497
Sorties, two shifts	2,559

Spares Short	Mon	Tues	Wed	Thurs	Fri	Sat	Sun	Total
One shift								
First wave	3	45	38	60	61	0	0	207
Second wave	0	0	0	0	0	0	0	0
Two shifts								
First wave	0	4	8	8	7	0	0	27
Second wave	0	0	0	0	0	0	0	0

able, but we see that this schedule would require overtime virtually every weekend, which is not sustainable. We would conclude from this example that the flying schedule tested here would not be sustainable with single-shift maintenance and would in fact require a second-shift maintenance operation to be feasible.

By varying the model inputs (daily flying schedule, initial pool of available aircraft, break rates, fix rates, and maintenance shift schedules), we can use this approach to estimate whether an ANG unit would be able to support the high sortie requirements implied by TFI UTE rate with a traditional single-shift maintenance operation or whether, in fact, a two-shift maintenance operation would be required.

Bibliography

Air Force Logistics Management Agency, *Analysis of Out-of-Hide Job Requirements Levied on Aircraft Maintenance Units,* Maxwell AFB, Gunter Annex, Ala., October 2005.

Albrecht, Mark J., *Labor Substitution in the Military Environment: Implications for Enlisted Force Management*, Santa Monica, Calif.: RAND Corporation, R-2330-MRAL, 1979. As of August 1, 2007:
http://www.rand.org/pubs/reports/R2330/

Amouzegar, Mahyar A., Lionel A. Galway, and Amanda B. Geller, *Supporting Expeditionary Aerospace Forces: Alternatives for Jet Engine Intermediate Maintenance*, Santa Monica, Calif.: RAND Corporation, MR-1431-AF, 2002. As of August 1, 2007:
http://www.rand.org/pubs/monograph_reports/MR1431/

Consolidated Manpower Database (CMDB), HQ Air Force Command Manpower Data System, U.S. Air Force, Directorate of Manpower, Organization, and Resources, September 30, 2004.

Dahlman, Carl J., Robert Kerchner, David E. Thaler, *Setting Requirements for Maintenance Manpower in the U.S. Air Force*, Santa Monica, Calif.: RAND Corporation, MR-1436-AF, 2002. As of August 1, 2007:
http://www.rand.org/pubs/monograph_reports/MR1436/

Feinberg, Amatzia, Hyman L. Shulman, Louis W. Miller, and Robert S. Tripp, *Supporting Expeditionary Aerospace Forces: Expanded Analysis of LANTIRN Options*, Santa Monica, Calif.: RAND Corporation, MR-1225-AF, 2001. As of August 1, 2007:
http://www.rand.org/pubs/monograph_reports/MR1225/

Geller, Amanda, David George, Robert S. Tripp, Mahyar A. Amouzegar, and Charles Robert Roll, Jr., *Supporting Air and Space Expeditionary Forces: Analysis of Maintenance Forward Support Location Operations*, Santa Monica, Calif.: RAND Corporation, MG-151-AF, 2004. As of August 1, 2007:
http://www.rand.org/pubs/monographs/MG151/

Lynch, Kristin F., John G. Drew, Sally Sleeper, William A. Williams, James M. Masters, Louis Luangkesorn, Robert S. Tripp, Dahlia S. Lichter, and Charles Robert Roll, Jr., *Supporting the Future Total Force: A Methodology for Evaluating Potential Air National Guard Mission Assignments*, Santa Monica, Calif.: RAND Corporation, MG-539-AF, 2007. As of October 1, 2007: http://www.rand.org/pubs/monographs/MG539/

Oliver, Steven A., *Cost and Valuation of Air Force Aircraft Maintenance Personnel Study,* Maxwell AFB, Gunter Annex, Ala.: Air Force Logistics Management Agency, August 2001.

Peltz, Eric, Hyman L. Shulman, Robert S. Tripp, Timothy Ramey, and John G. Drew, *Supporting Expeditionary Aerospace Forces: An Analysis of F-15 Avionics Options*, Santa Monica, Calif.: RAND Corporation, MR-1174-AF, 2001. As of August 1, 2007:
http://www.rand.org/pubs/monograph_reports/MR1174/

Robbert, Albert A., William A. Williams, and Cynthia R. Cook, *Principles for Determining the Air Force Active/Reserve Mix*, Santa Monica, Calif.: RAND Corporation, MR-1091-AF, 1999. As of August 1, 2007:
http://www.rand.org/pubs/monograph_reports/MR1091/

Taylor, William W., S. Craig Moore, and Charles Robert Roll, Jr., *The Air Force Pilot Shortage: A Crisis for Operational Units*? Santa Monica, Calif.: RAND Corporation, MR-1204-AF, 2000. As of August 1, 2007:
http://www.rand.org/pubs/monograph_reports/MR1204/

Tripp, Robert S., Kristin F. Lynch, Ronald G. McGarvey, Don Snyder, Raymond A. Pyles, William A. Williams, and Charles Robert Roll, Jr., *Strategic Analysis of Air National Guard Combat Support and Reachback Functions*, Santa Monica, Calif.: RAND Corporation, MG-375-AF, 2006. As of August 1, 2007:
http://www.rand.org/pubs/monographs/MG375/

U.S. Department of Defense, *Quadrennial Defense Review Report,* September 30, 2001.

———, "The United States Military Enlisted Rank Insignia," Web page, March 31, 2004. As of August 14, 2007:
http://www.defenselink.mil/specials/insignias/enlisted.html

———, "BRAC Commission Actions," briefing, September 1, 2005.

U.S. Air Force, *Air Force Basic Doctrine*, Document 1, November 17, 2003a.

———, *Air Force Instruction 38-201*, December 30, 2003b.

———, *Air Force Instruction 21-101, Aerospace Equipment Maintenance Management,* June 2004.

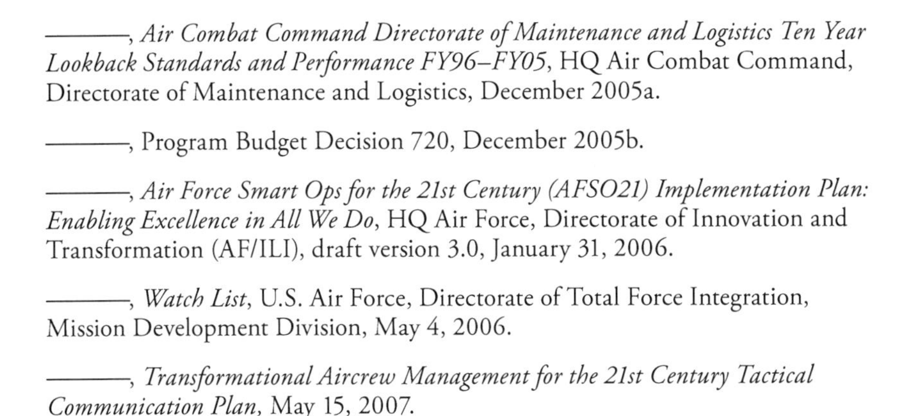

———, *Air Combat Command Directorate of Maintenance and Logistics Ten Year Lookback Standards and Performance FY96–FY05*, HQ Air Combat Command, Directorate of Maintenance and Logistics, December 2005a.

———, Program Budget Decision 720, December 2005b.

———, *Air Force Smart Ops for the 21st Century (AFSO21) Implementation Plan: Enabling Excellence in All We Do*, HQ Air Force, Directorate of Innovation and Transformation (AF/ILI), draft version 3.0, January 31, 2006.

———, *Watch List*, U.S. Air Force, Directorate of Total Force Integration, Mission Development Division, May 4, 2006.

———, *Transformational Aircrew Management for the 21st Century Tactical Communication Plan*, May 15, 2007.